Stealing Cars

STEALING CARS

Technology & Society from the Model T to the Gran Torino

JOHN A. HEITMANN
& REBECCA H. MORALES

Johns Hopkins University Press
Baltimore

Johns Hopkins University Press
2715 North Charles Street
Baltimore, Maryland 21218-4363
www.press.jhu.edu

Library of Congress Cataloging-in-Publication Data

Heitmann, John Alfred.
   Stealing cars : technology and society from the Model T to the Gran Torino /
John A. Heitmann and Rebecca H. Morales.
      pages cm
   Includes bibliographical references and index.
   ISBN 978-1-4214-1297-9 (hardcover : alk. paper) — ISBN 978-1-4214-1298-6
(electronic) — ISBN 1-4214-1297-7 (hardcover : alk. paper) — ISBN 1-4214-1298-5
(electronic)
      1. Automobile theft—United States—History. 2. Automobile theft—United
States—Prevention. 3. Automobiles—Technological innovations. 4. Automobile
thieves—United States. 5. Grand Theft Auto games—Social aspects. 6. Automobile
theft—Mexican-American Border Region. I. Morales, Rebecca. II. Title.
      HV6658.H45 2014
      364.16'286292220973—dc23   2013032111

A catalog record for this book is available from the British Library.

Special discounts are available for bulk purchases of this book. For more informa-
tion, please contact Special Sales at 410-516-6936 or specialsales@press.jhu.edu.

Johns Hopkins University Press uses environmentally friendly book materials,
including recycled text paper that is composed of at least 30 percent post-
consumer waste, whenever possible.

Contents

Acknowledgments

Rarely is a book of this complexity the result of authors working without considerable assistance. For author Heitmann, the list of those who contributed to the work is long, and there is always concern that someone's name will be left out. Certainly former student Peter Cajka, now at Boston College, was crucial to getting this work off the ground, as his research formed the core of what was initially included on the history of auto theft in *The Automobile and American Life*. Later we traveled to the National Archives in College Park, Maryland, where more source mining took place. Peter also did considerable work on the topic of electronic gaming, which is a part of chapter 6.

During the fall of 2010, a meeting with Johns Hopkins University Press editor Ashleigh McKown at the SHOT Meeting in Tacoma encouraged me to move forward on the project. When Ashleigh subsequently moved on, senior editor Robert Brugger shepherded the study, making sure that the broader significance of car theft, including motives, was examined. Two anonymous reviewers recommended revisions to bring this work up to the standard of excellence that the Press is known for. Finally, copy editor Lois Crum did a remarkable job in pointing out inconsistencies and smoothening prose.

What proved decisive, however, was my appointment as the visiting Knapp Chair in the Liberal Arts at the University of San Diego, where Dean of the College of Arts and Sciences Mary Boyd, History Department Chair Ken Serbin, and Professor Molly McClain gave me the opportunity to work without distraction on the topic. Upon my return to the University of Dayton, Dean of the College of Arts and Sciences Paul Benson and History Department Chairs Julius Amin and Juan Santamarina ensured that I had the time and resources to finish the job. A summer University of Dayton Research Council Fellowship provided generous support during the summer of 2011. Todd Uhlman, a colleague in the history department, taught me much about film analysis and forced me to think philosophically, culturally, and above all abstractly. His thinking left a lasting imprint on the cultural analysis contained in this book. Finally, Roger Morris, Vice President of Communications at the National Insurance Crime Bureau, welcomed me to the organization's headquarters in Chicago, where he freely shared historical material that is seen throughout this book.

For my *Automobile and American Life,* editor Niki Johnson, now retired from the University of Dayton Research Institute, tirelessly polished my prose and caught errors and omissions. Afterward, I heard many compliments concerning the quality of that book, which was the result of Niki's keen eyes and not my more casual attitude to precision. Again Niki stepped forward to edit this manuscript, and I am forever grateful for her unselfish commitment to helping me.

People closest to me also proved crucial to the completion of this book project. Coauthor Rebecca H. Morales willingly gave her time and prodded, cajoled, and encouraged as needed. And then my family had its role as well. My wife Kaye manned the fort in Dayton while I was in sunny San Diego, with the small consolation of operating a new snow blower while I was playing tennis at Balboa Park! Daughter Lisa, living in San Diego, provided meals and company, along with son-in-law Tony, whose car-restoration projects kept my feet on the ground when I was inclined to just read books on automotive history.

I dedicate this book to my Johns Hopkins University dissertation director, the late Owen Hannaway. Back in the 1980s, Owen took this pretty rough would-be scholar and made him much better, never forcefully dictating a course of study, but always gently directing. I can still see him beaming when he talked about the docks of Glasgow! With a totally open mind toward the choice of subject material and approach, Owen understood that in a subtle way, an author must intimately connect with the material, making it his own.

For Rebecca Morales, this was a journey full of surprises, with potholes, caveats, and unexpected insights along the way. Perhaps the single person who proved to be the most important touchstone was Frank G. Scafidi, Director of Public Affairs for the National Insurance Crime Bureau in Sacramento, California. Without fail, Frank provided a reality check whenever one was needed, and he opened doors to hard-to-access data and people.

When it came to two boots on the ground, Steve Witte, Chula Vista Police Sergeant and Head of the San Diego Regional Auto Theft Task

Force, and Richard Valdemar, a retired Sergeant with the Los Angeles County Sheriff's Department, were invaluable. Their knowledge provided the nuance that is lost in mere numbers on a page; their interpretation of events shed light on obscure information.

The long and at times difficult relations between Mexico and the United States over auto theft were made more understandable by David A. Shirk, Associate Professor of Political Science and International Relations and Director of the Trans-Border Institute at the University of San Diego. Further assistance was provided by Efrain Aceves, a scholar and policy analyst in Mexico City who shared his broad expertise that spans many disciplines.

In addition, I would like to thank the following people for their generous time and thoughts: Annette Villarreal, Supervisor, Texas Department of Public Safety, Border Auto Theft Information Center / CID, El Paso, Texas; Ralph Lumpkin, Director of Operations for Area 3, National Insurance Crime Bureau, Chula Vista, California; and Dennis Frias, California Operations Manager, Oficina Coordinadora de Riesgos Asegurados S.C., Covina, California.

And last, but not least, was the intangible support provided by my family and friends. Thank you.

Stealing Cars

Park at Your Own Risk

What the hell does everybody want with my Gran Torino?
WALT KOWALSKI [CLINT EASTWOOD], *GRAN TORINO* (2008)

Automobile theft is a crime at the margins of American life. Yet it also reflects themes that are at the core of both modern existence and what it means to be human. For the thief, the act can be a vicarious experience. It is a moment that linguist Jeffrey T. Schnapp suggests vaults the perpetrator into "the world as its conqueror, ruler and judge." In a classic role reversal, the often clever and technologically adept thief gains freedom at the expense of an "unhorsed" owner, who has lost autonomy and identity. The criminal, who usually comes from the periphery of society, moves, albeit temporarily, into a life "of bigger living," a world in which class distinctions and material possessions have been temporarily suspended.[1]

Auto theft is such a common occurrence in America that we hardly take notice when it happens—unless the car is ours. Yet insurance industry statistics tell us that in 2013 someone steals a car every thirty-three seconds. If we were to string out the annual total of stolen cars bumper to bumper, the line would stretch from New York City to Phoenix, Arizona. Auto theft may not be central to our everyday lives, but it is far from inconsequential, particularly when it relates directly to more serious crime.

Many Americans have directly or indirectly experienced the theft of a car, our most prized possession after our home. Our personal experiences, however, capture only a portion of the complexity and changing nature of auto theft in the United States from the early days to the present. Several questions stand out: Who steals cars, and why? How has the crime changed over time? Is car theft motivated by the drug of speed and thrills, sovereign individualism, easy money, wanting what one does not have, race and class antagonisms, the need for transportation, repressed sexual impulses, boredom, or something else? Or is it, as anthropologist Sarah S. Lochlann Jain suggests, almost lauditorily where "freedom meets regulation and a potential for individuation rubs uneasily against actualized homogeneity"?[2] On the flip side, why did so many Americans up to and through the 1960s leave their keys in their cars, purportedly objects that were loved and often considered part of the family? When we examine the design of automobiles, there are questions concerning what antitheft measures were incorporated into the cars coming off the assembly line and why, and about the thousands of inventors' aftermarket technologies. Automobile theft has presented opportunities for inventors to create devices to thwart thieves; thieves then use creative means to overcome the ingenious locks and electronic alarms. And with increasingly sophisticated technology and approaches to theft have come issues regarding the who, what, and why of institutional responses. From the perspective of society and the built environment, the question of why some places are hot spots for auto theft and others are relatively safe begs to be explored. And in a transnational age, it is pertinent to ask how international forces work to the advantage of thieves.

But why take time to tell this history at the margins? The narrative is not about luscious cars or creative engineers and businessmen but about everyday people, both lawbreakers and victims. Fundamentally, however, it is a telling account about a significant slice of the American past. The topic also fits well with the recent trend in historiography relating to the automobile; the focus in that field seems to be moving away from producers and toward users, even if the users happen to be thieves and joyriders. And it illustrates just how central the automobile has been to American life, beginning in the twentieth century.

Although the meaning of auto theft in the history of twentieth-century American life remains somewhat unclear, one simple conclusion is that the auto thief steals an owner's freedom, both literally and figuratively. Movement is transformed into immobility, and vice versa. As the Brazilian literary scholar Guillermo Giucci has argued in a different context, the deed marks "the demise of an illusion and the loss of the hope of salvation through acceleration." According to the English sociologist John Urry, this sacred thing called the car is central to the modernization of urban life, including life's disappointments. But movement, or kinetic modernity, cannot be understood, says Urry, without the conceptual mirror-image twins flexibility and coercion.[3] Indeed, the history of auto theft links intimately to the interplay of these notions, and some of those automotive users—thieves unlucky enough to be caught—end up with the ultimate loss of freedom, being sentenced to jail or prison.

Thus, the history of automobile theft in twentieth-century America bridges science, psychology, economics, technology, and society. As such, it helps one explore the foundations of criminal motives, techniques, and organization; the development of a variety of antitheft technological countermeasures; the rise of institutional rejoinders from government, the insurance industry, and manufacturers; the environmental solutions created by city planners and architects; and opportunities, challenges, and diplomatic and legal relations between nations in an international society. Furthermore, in the history of auto theft one can see recurring cycles. During every era authorities have proclaimed that auto theft was largely solved. However, new criminal strategies would thwart the best of efforts, and the problem would become bigger than ever. Only in the recent past have we experienced a statistical decline in this criminal activity.

Curiously, even though victims who feel personally violated abound and the cost of auto theft to Americans remains sky high, literature and film frequently lionize the auto thief. The act often seems to be victimless—as long as the owner has insurance—and in cultural representations the professional car thief appears as a clever hero, satisfying personal urges that reflect the central values of traditional American car culture: masculinity, status, and freedom. As long as it is not our car, the bad guys are not so bad.

For generations, historians studying the automobile in America have concentrated their efforts on themes commonly associated with the history of technology or business.[4] These scholars have also been decidedly "American-centric"; that is, their work has rarely crossed

national or continental boundaries. Traditionally, automobile history has described the cars themselves or individuals associated with
them, rarely offering either broad or interpretative conclusions. During
the past decade, however, a "new" automotive history has gradually
emerged, one that explores users rather than producers and in so doing
delves into various interstices that include automobile operation, re-
pair, tinkering, and safety, and the connection between automobiles
and the environment. And crime, more specifically in this case auto
theft, has its own niche that also reflects a much larger scene. This
study continues the work of contemporary scholars who look at history from the bottom up yet also focus on the importance of nations,
government, and culture in mediating the relationship between technology and society.

The plan of this book is mostly chronological. It begins with an examination of American auto theft from the early era to the onset of
World War II. During that period several recurrent themes appeared:
high incidences of thefts; joyriders taking the wheel for the thrills or
convenience of travel; hardened professional criminals in it for money;
governments expanding at all levels to preserve order; inventors devis-
ing apparatuses to stymie the thieves; and technology appearing to be
triumphant, but only for a moment. Later, around 1980, the problem of
theft became so significant that authorities and manufacturers stepped
up both legal sanctions and technological security devices. By means
of walls, gates, and cameras, designers and planners tried to devise
"defensible spaces" that would protect cars and people from criminals. In spite of these efforts, auto theft if anything climbed during the
1980s. Technology once again appeared to solve this social problem,
now with digital electronic security measures. An entire chapter dis-
cusses the digital age—the evolution of the automobile as an electronic
system—and its long-term effect on patterns of auto theft. Because an
increasingly caffeinated, digitized, and interconnected age has com-
pressed space and made borders more porous than ever, car theft has
become a major problem along the U.S.-Mexico border. Gangs, drugs,
and cars have made California the hot spot for auto theft, displacing
both Detroit and Newark, New Jersey.

Auto theft has rarely been an isolated illegal activity. During the 1920s it was intimately tied to Prohibition, in the 1930s to bank robberies, and in more recent times to gangs and drugs, and perhaps terrorism. *Stealing Cars* ends with a survey of the recent past, tracing automobile theft in an age when the numbers of stolen cars has declined but criminal methods grow vastly more sophisticated and increasingly are based on computer and electromagnetic technologies. Auto theft has always been a game, intellectually and in terms of technique, but only recently has it become an activity simulated by electronic games that are played by millions of young adults in their twenties and thirties, as well as children after school.

While this study centers on the American scene, it makes no claim of American exceptionalism. Like so many other aspects of modern life that have been globalized, automobile theft has increasingly become a complex transnational issue, with important localized differences tied to opportunity, incentives, political systems, local policing, and culture.[5] As long as the automobile remains an object of desire and the rhetoric of freedom collides with hyperregulation, it is doubtful that the problem of auto theft will be totally solved, no matter what deterrent technologies are introduced. Human beings always seem to be clever enough to circumvent the most sophisticated of antitheft devices. After all, the human spirit—both legally and illegally—thrives on overcoming challenges that act to separate, regulate, and restrain it.

John Heitmann first uncovered the historical topic of auto theft when piecing together sections of what became *The Automobile and American Life* (McFarland, 2009). He was astonished at the scale of pre–World War II auto theft, fascinated with the characters involved, and captivated by a host of deterrent technologies that were introduced to supposedly solve the problem. It was an unlikely story about cars, antitheft devices, teenagers, hardened criminals, the police, the insurance industry, and J. Edgar Hoover. Rebecca Morales, with her extensive knowledge of the international automobile industry, in Latin America and Mexico especially, minorities in the United States, and the built environment, came in later to round out the picture.

Both authors directly or indirectly have experienced the theft of a car. The first time this occurred in Heitmann's life was in 1980 when someone tried unsuccessfully to steal his green 1973 Ford Pinto from the Pimlico Park and Ride in Baltimore. To this day he wonders why anyone would want to take that car, especially since by then the word was out that the Pinto's gas tank, upon rear impact, tended to explode and fry the car's occupants as the doors jammed shut. But the thief was thwarted because he did not stick the screwdriver deep enough into the mechanism before trying to force it to unlock. The second occasion involved his 1979 Malibu Classic, parked at a Sears Hardware Store near his home in Centerville, Ohio. Leaving the store after purchasing a fastener, he was surprised to find a swarthy, curly-headed young man trying to start the car! Heitmann walked up to the scene and heard the culprit quickly explain that he had thought he was in a different, identical car. A very unlikely story, to be sure, but, in shock, Heitmann allowed the quick-witted would-be thief to walk away.

Rebecca Morales has a different set of stories involving auto theft. She had her 1991 Acura Integra stolen three times in front of her home in San Diego—the first two times by joyriders and the last time by professionals who left behind a stripped carcass. In a twist of irony, she believes the car she obtained to replace her dismantled car may have been stolen and subjected to what the FBI calls "cloning," or creating a "new" car from stolen parts—a process that is becoming increasingly common with the growing internationalization of auto theft.

These personal experiences capture only a portion of the complexity and changing nature of auto theft in the United States from the early days to the present. Together, the authors explore how auto theft has occurred in various periods, as well as why, by whom, and what have been the responses to it. It is a history of technology but also of a society that is continuously assaulted by lawbreakers and constantly regrouping to meet the challenge. At the same time, we Americans, surely influenced by our culture, are conflicted over whether to demonize the thief or applaud his ingenuity and courage.

"Stop, Thief!"

Not only is the motor vehicle a particularly valuable piece of property . . . but it furnishes at the same time an almost ideal getaway. . . . With the automobile there is no planning to be done. With a thousand divergent roads open to him and a vehicle possessing almost unlimited speed, escape is practically automatic.

COUNTRY LIFE, **1919**

How can we who own automobiles feel safe in keeping one when we have here in this fair city of ours [Chicago] places to dispose of them so readily. Anyone wanting a set of wheels & tires or other accessories which are stripped from cars stolen in Indiana, Mich or other states no doubt but mostly local that is [*sic*] stolen here can get same very cheap. Our local police do not seem to prevent it. So many of our large automobile parts stores seem to have protection to handle such goods. Can something not be done about it?

LETTER TO DEPARTMENT OF JUSTICE, 1932

The automobile was a primary object for thieves and a perfect accessory to crime during the twentieth century. Not knowing what the future held, however, in 1901 a physician and early steamer owner was confident that the coming of the automobile would substantially decrease personal-transportation thievery: "When I leave my machine at the door of a patient's house I am sure to find it there on my return. Not always so with the horse: he may have skipped off as the result of a flying paper or the uncouth yell of a street gamin, and the expense of broken harness, wagon, and probably worse has to be met."[1] That solitary impression soon proved to be wrong. The *Horseless Age* first reported an auto theft during the fall of 1902, although there seemed to be some contention as to when the theft first took place: "H. Clark Saunders, New Brunswick, N.J. writes us that the theft of an automobile mentioned in our last issue [1903] was not the first on record, as Mr. Laurenz Schmalholz, New Brunswick, N.J., while in Trenton, N.J. on the day before the election last November [1902], had his Pierce Motorette stolen from the stables of the United States Hotel, and it was not until late the following day that the machine was found."[2]

The security of the steamer in particular was noted by early automotive pioneers, who clearly recognized that their vehicles were far from secure. A 1901 article on locking devices for cars stated: "It is not a safe proceeding to let an automobile stand in the street so that the operation of a conspicuous hand lever will start the vehicle; and it may be said that we are approaching a period where it will not be safe to let a vehicle stand in the street which can be started by any person thoroughly familiar with that particular machine, but without the necessary key or keys. Most manufacturers are recognizing these conditions and are providing means either against accidental starting or both against this and malicious designs."[3]

Apparently, steam and electric vehicles were better equipped with locks than their internal-combustion-powered rivals of that era. Steam-powered vehicles typically had locks that immobilized the throttle, "thereby preventing any possible interference by the over curious meddler."[4] For example, the Victor Steam Carriage Company used a particularly complex device that nevertheless did not ensure security:

> A spring actuated catch is employed, which locks the throttle lever, inside the seat, is fastened to a double-armed lever, which occupies a near horizontal position when the throttle lever is in the off position. Inside the seat there is a single-armed lever, the lower part of which stands nearly vertical normally and the upper part which is inclined forwardly. A catch on this lever engages with the one arm of the double-armed lever fastened to the throttle lever shaft and prevents the motion of the latter. The single-armed lever is held in position by a coiled spring. The upper end of the lever bears against the seatboard, and the spring is sufficiently powerful to lift the board when nobody is sitting on it. When anybody sits

down when the seatboard is depressed, the spring is extended and the catch released.[5]

Electric vehicles typically came with a key that activated a switch so that with one turn, the current to the motor was cut off. Yet the designers of many of the earliest vehicles powered by internal-combustion engines were cavalier about security. For example, the Winton used a common snap switch mounted to a porcelain base. The owner could remove the hard rubber handle when the vehicle was not in use, but this was not usually done, since the switch could be operated without it. Packards featured two push-button switches that interrupted the igniter circuit. Ordinarily, one of these switches was placed on the kneeboard and the other in the battery box, which could be locked. The 1901 Hayes-Apperson used a switch that the company itself had constructed. The *Horseless Age* reported, "The contact pieces form arcs of a circle, and over these moves a double-armed flat contact lever swiveling in the center of the circle and held down to contact pieces by a shoulder butterfly screw. This screw can easily be removed and the lever carried along."[6] But if a would-be thief possessed a lever from one Hayes-Apperson, that lever could be used to steal any other. French vehicles were hardly better in terms of theft-proof design. In a De Dion-Bouton, a slightly tapered plug was used to open the igniter circuit when turning off the car; this plug could be carried along in the pocket of a dismounted driver. The button from any De Dion-Bouton would work. But who would suspect that anyone from the car-owning aristocracy of the day would covet his neighbor's horseless carriage? The motives of the often suspect chauffeur or garage owner were a different matter, however.[7]

Aftermarket manufacturers and independent inventors soon got busy developing devices to satisfy the insecurities of a growing number of automobile owners. By 1903 the Auto Lock Company of Chicago began advertising its Oldsmobile Lock, claiming, "It locks everything (even while the motor is running), prevents theft and meddling. Once used, always used."[8] In July 1909, Orville M. Tustison of Bainbridge, Indiana, patented his "Circuit-Closer," which employed a Yale-type lock with mechanical and electrical mechanisms that ultimately served as a spark coil kill switch. Located prominently on the dashboard, Tustison's device was only as good as the Yale lock and box that housed the device.[9]

These and other technological devices were also limited by the effectiveness of the local law enforcement of the day. Beginning around 1910, police procedure evolved only gradually and, one might surmise, haphazardly, as the car-theft problem became increasingly acute. In the best-organized police departments, the report of a stolen car was documented by an entry into the department's log book.[10] Since there was little, if any, communication between police departments in those early days, standard procedures for exchanging information did not exist, and only rarely were theft reports transmitted to other regions. Therefore, it was normally left up to insurers or vehicle owners to publish and disseminate a reward offer and get the word out that a particular vehicle had been stolen.

The insurance companies took the lead early on. During the summer of 1912, eleven insurance companies formed the Automobile Protective & Information Bureau, with the purpose of disseminating information about specific stolen vehicles.[11] Rewards ranged from twenty-five dollars to five hundred dollars, and notices were often printed on 8×10 inch manila cards and mailed to nearby police departments. A woodcut, obtained from a car dealer who had used the image for advertising purposes, was stamped on the "wanted poster" along with such details as the vehicle's color, size of tires, type of headlamps, and whether the vehicle had a windshield. Given the poor roads and the durability of early cars, stolen vehicles were rarely taken farther away than 150 miles, and thus mailings were confined to the hinterlands near where the theft had occurred. It was assumed, accurately or not, that police would do their honest best to recover the vehicle.

One can only imagine the consternation of car owners when they learned that New York City police were neglecting cases of auto theft.[12] In 1914, according to the district attorney's office, some assigned detectives made no effort to pursue the criminals until insurance companies offered rewards. With that incentive, the New York police garnered an extra ten thousand to fifteen thousand dollars when they arrested twelve thieves and recovered twenty stolen cars. The reward system also led to breaking the first auto theft ring on record. John Gargare, owner of a Lakewood, New Jersey, garage, was indicted on six counts of auto theft in 1914 by the same New York City district attorney. Gargare and his accomplices specialized in Packards and Pierce-Arrows, demonstrating a preference for high-end vehicles that future professional auto thieves often imitated.

The failure of local law enforcement to stem the growing tide of thefts prompted the insurance industry to take the lead. The private sector has remained active to this day in both gathering information on the crime and tracking down the crooks and the cars. At the beginning of the automobile age, insurance policies typically covered only fire and damage. A 1910 article in the *New York Times* commented: "A great many automobilists do not give enough consideration to the theft clause of the floating fire policy. Cars are stolen quite frequently, and it is seldom that the fire insurance companies are able to trace the cars[;] consequently they are compelled to pay a total loss under the policy."[13]

Losses became so great that two years later, in 1912, the National Automobile Theft Bureau (NATB) was established as an arm of the National Automobile Underwriters Conference. Supported by member insurance companies, the NATB served as a private police force and a nationwide information bureau that overlapped with governmental authorities.[14] NATB personnel trained police officers and encouraged the standardization of stolen-car information. And contrary to the assertions that the auto industry neglected the auto-theft problem, in a contentious reorganization of the NATB in 1926, A. C. Anderson, General Motors's general comptroller, forcefully brokered the unification of regional insurance interests into what emerged as national agency.[15]

The relationship between the NATB and local police was often tenuous, since the boundary between private and public was being crossed. With a wealth of expertise in matters related to automobile identification and the methods of criminals, NATB agents educated local police who had little direct knowledge in these matters, and the NATB was active in the establishment of dedicated auto-theft investigative units within police departments.[16] Yet just as the police from time to time did not escape charges of lack of motivation and corruption, insurance personnel also were subject to the temptation to personally profit from stolen cars. For example, in 1914 a chauffeur and an insurance adjuster, working together, were accused of making a small fortune in the business of hot cars.[17]

MASS PRODUCTION AND THE POST–WORLD WAR I RISE IN AUTO THEFT

With Ford's "democratization" of the automobile and an explosion in the number of vehicles came an epidemic of automobile theft. Machines produced in mass quantities made easy prey for joyriders, common thieves, and skilled, organized professional criminals. Moreover, the automobile was valuable and mobile, and its parts were often interchangeable. Domestic and international markets for stolen automobiles and parts yielded high profits and relatively low risk. Interchangeable parts also enabled thieves to quickly reconstruct and disguise stolen automobiles. As evinced by the ability of thieves to alter serial numbers, duplicate registration papers, switch radiators, and replace entire engine blocks, Fordism's inherent uniformity welcomed theft. With few exceptions, thieves sought out and stole the most ubiquitous automobiles; popular, mid-priced models were most likely to be stolen, along with the easy-to-steal Model T.

Until the introduction of the electric self-starter in 1912, automobiles employed a battery-magneto switch along with a crank.[18] The automobilist turned the switch to B (battery), got outside the car, and cranked the engine; once it started, he moved the lever to M (magneto) and adjusted the carburetor. On the early Ford Model T, the battery-magneto switch had a brass lever key, but there were only two types, with either a round or a square shank. Later, in 1919, Ford offered an optional lockable electric starter, but the company used only twenty-four key patterns. To make things easy for the thief, each code was stamped on both the key and the starter plate.

Most significantly, however, the very nature and scale of criminality was transformed by automobility. Unlike other stolen goods, the automobile enabled its own escape. One such episode happened in 1925,

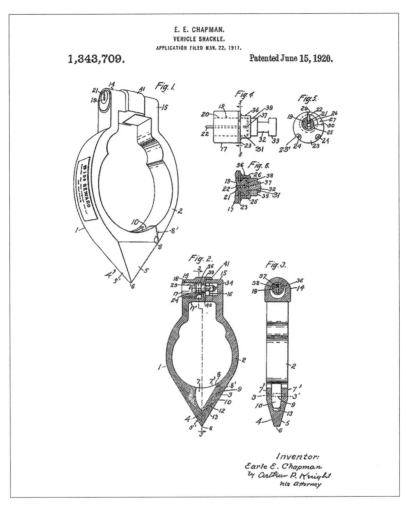

E. E. CHAPMAN.
VEHICLE SHACKLE.
APPLICATION FILED MAR. 22, 1917.

1,343,709.

Patented June 15, 1920.

Inventor:
Earle E. Chapman
by Arthur P. Knight
his Attorney

Between 1914 and 1925 there were at least 25 patents related to a wheel chock or wheel lock that shackled a wood spoke wheel. The Miller-Chapman Security Auto-Theft Signal System came in 32 sizes to fit different rims and tires. A would-be thief could defeat the device by simply removing the wheel. The Miller-Chapman patent of 1920 served as the basis of a Japanese inventor's patent in the 1990s that led to the manufacture of an antitheft wheel chock currently available for purchase on the Internet.

when five men held up a cashier and timekeeper at a construction site in the Bronx, took two thousand dollars, and then fled in the victim's car. The *New York Times* reported that the thugs, "as they fled . . . fired a shot from the automobile at a number of workmen who had dropped their tools and were giving chase. The robbers' car was out of sight when they reached 165th Street and Jerome Avenue."[19]

In 1916 a New York police official was stating the obvious when he commented, "The automobile is a very easy thing to steal and a hard thing to find." As early as 1915, 401 automobiles were stolen in New York and only 338 were recovered.[20] By 1920, it was estimated that one-tenth of the cars manufactured annually were stolen.[21] Astonishingly, in 1925 it was estimated that 200,000 to 250,000 cars were stolen annually. The automobile age had ushered in a new era of crime and a new type of criminal, the joyrider.[22]

This crime wave could not be attributed to just one kind of criminal, though, particularly one who considered the act to be a casual "borrowing" of a vehicle. Automobile theft added new categories of crimes, and the motor vehicle played a central role in burglary and housebreaking. In response, police began using automobiles to patrol. In 1922 Chicago police complained that their worn-out "tin lizzies" should be scrapped; they could not catch the high-powered holdup car that traveled at sixty miles an hour.[23] Even with the growth of government and the advent of patrolling, police forces were outmaneuvered by mobile criminals. Contrary to the iconic Prohibition image of police forces smashing barrels of alcohol, municipal police forces may have dealt with stolen automobiles on a more regular basis. Automobile theft developed as a complex phenomenon, one that was not easily characterized in terms of motives or methods. It became as complex as American life in the machine age. In Philadelphia in 1925, 8,896 people were arrested for assault and battery by the automobile, used as a weapon.[24]

Crime expert George C. Henderson, in his popular *Keys to Crookdom* (1924), placed car thieves in five categories: commercial thieves and hardened criminals; strippers; traveling crooks; robbers or bandits; and "Joy riders, kids, imbeciles, dope fiends, incorrigibles, roughnecks and members of youthful gangs [who] steal cars just to ride around town."[25] As to the "why" of youthful offenders, W. S. Jennings, commenting on those incarcerated in Indiana's Jeffersonville Reformatory, asserted that "divorces, broken homes, children spoiled in raising by neglectful parents, or by equally neglected over-indulgent ones; the absence of rational home life to counteract city temptations; [and]

failure to learn self control in early life" were the reasons adolescents were breaking the law.[26]

In reality, there were almost as many reasons for becoming an auto thief as there were thieves. One purported auto thief, writing a confession in a 1925 issue of *Your Car: A Magazine of Romance, Fact, and Fiction,* explained that he "drifted into stealing automobiles, because it seemed the easiest way to get what was to me a lot of money, quickly. I had a champagne appetite and a dishwasher income." Beginning with the theft of automobile jacks, tire irons, and tires, this repentant criminal organized a gang of three, including one skilled mechanic who had "graduated from one of 'Detroit's finest factories.'" The trio, careful to study the daily habits of the car owners under consideration, concentrated on stealing Buicks in New York City and then moving them to a shop in Westchester, New York. A chance apprehension interrupted their practice, and the writer was determined to go straight after serving a two-year prison sentence. He cautioned owners by highlighting a strategy that remains viable to this day: "The stories I read about automobile thieves and how slick they are, opening any lock in fifteen minutes, installing wiring systems of their own and all that sort of stuff, make me laugh. Why should a thief go to all that trouble when right around the corner he can find another car without any locks on it, except the ignition lock, which his master key will open as quickly as the owners?"[27]

In sum, auto theft was often one of several interrelated crimes perpetrated by lawbreakers. The automobile created new opportunities for criminals of all persuasions and consequently confronted legal authorities with a myriad of problems. One author noted, "As automobile thefts increase burglaries and robberies increase."[28] The automobile itself was stolen, but the automobile also played a central role in kidnapping, rum running, larceny, burglary, traffic crimes, robberies, and deadly accidents of the "lawless years."[29]

AUTOMOBILE THEFT AND THE 1920s CITY SCENE

By the 1920s, automobile theft was most acute in Detroit and Los Angeles. "Naturally Detroit is peculiarly liable to this trouble because it has such a large floating population of men trained to mechanical expertise in the various factories."[30] It stood to reason that Ford's workers stole Ford's cars. Arthur Evans Wood reported that in 1928 in Detroit a total of 11,259 cars were stolen.[31] Of those thefts, less than 10 percent led to an eventual arrest, and only 50 percent of that group were ever prosecuted. In the end, only 25 percent of the persons arrested for auto theft in this particular group were ever convicted. Since at that time many thieves ended up paying off "coppers" to avoid apprehension, one might conclude that this crime actually did pay.

The same year, 10,813 automobiles were stolen in Los Angeles. By the 1920s, Los Angeles had the most automobiles per resident in the United States. This fact was clearly changing the face of crime in the City of Angels. As historian Scott Bottles points out, "By 1925, every other Angelino owned an automobile as opposed to the rest of the country where there was only one car for every six people."[32] Angelinos had more opportunities to steal cars, and some took those opportunities. In 1916 some 1,300 cars were stolen and 85 percent were recovered; a decade later more than 10,000 were taken with an 89 percent recovery rate.[33] Theft statistics remained in the range of 5,000 to 8,000 cars per year up to the onset of World War II.

Baltimore, New York City, Rochester, Buffalo, Cleveland, Omaha, St. Louis, and many other cities also experienced major problems related to automobile theft. In 1918, the year before federal legislation was enacted to deal with the problem, Chicago experienced more than 2,600 thefts; St. Louis, 2,241; Kansas City, 1,144; and Cleveland, 2,076. However, in an article published in *Country Life,* Alexander Johnson revealed that the problem existed outside urban America, too: "We who live in the country are not quite as subject as our urban brethren to this abominable outrage, but automobile stealing is carried on even in the rural districts."[34] Although cars from the countryside certainly were stolen from time to time, auto theft has remained largely an urban issue. Joyriders can be found in every locale; gangs, rings, bootleggers, and drugs were and are very much a part of the city scene.

Of course, within each city some neighborhoods were more secure than others. And there were "hot spots," some of them connected with the race of the majority of their inhabitants. Such was the case in Chicago in the early 1930s, where in several redlined districts insurance underwriters refused to issue policies "except under special arrangement." In these African American neighborhoods "conditions [were] so deplorable . . . that resident motorists [could] not obtain any insurance." To clarify why it made such a decision, the insurance industry stated, "The reason that Negros cannot secure theft insurance is not one of discrimination, but more or less one of character. Color does not play any part."[35]

THE ONUS ON OWNERS

Despite the obvious part played by misguided youth and bad guys, one theme—owner negligence—emerged early in the twentieth century and undoubtedly took pressure off manufacturers for their perceived lack of interest in incorporating theft deterrents in many of their vehicles. A 1916 insurance company pamphlet entitled "Emergency Instructions" warned owners that "when dining in a public restaurant the driver of the car should be seated in such a position that he can observe his car." Basic instructions also cautioned the owner, "[Do] not leave your car unprotected on the street or any place at any time."[36] However, one source reported that in 1922 many automobile owners left keys in their unlocked cars. An article in *Popular Mechanics Magazine* observed: "Approximately seventy-five percent of all the cars that were stolen were not locked at all."[37] The prevailing attitude of the day was that automobile theft was usually the owner's fault.

In 1929 E. L. Rickards, manager of the Automobile Protective and Information Bureau in Chicago, stated: "A man or woman who leaves his car unlocked and unattended is committing an offense against society." Some early automobile thefts were performed by owners who would "steal their own car." To collect on insurance, owners would strip the car of accessories and move it to an out-of-the-way location, then file an insurance claim. "An auto, for instance, that is insured for $2,000 is reported by the owner as having been stolen. The machine is worth $1,500. So the owner, collecting his theft insurance, makes a clean profit of $500."[38]

Owners in debt often defrauded insurance companies as well. One source reported in 1917 that "an automobile owner, after using his insured car for nine or ten months, discovers that its market value is 40 percent lower than when first purchased; also the cost of maintaining the machine, oil, gasoline, tires, repairs, etc., is considerably in excess of the figure on which his first maintenance costs were based."[39]

One could go overboard in putting the onus on owners, however. For example, in a 1917 magazine article, drivers were chastised for leaving their automobiles unattended for an hour or more.

TECHNOLOGICAL COUNTERMEASURES

Beyond commonsense precautions, automobile owners were advised to take preventive measures to stop early car thieves. Perhaps the most bizarre was the Bosco "Collapsible Rubber Driver," which was made in Akron, Ohio. Ad copy for the rubber man claimed, "Locks may be picked or jimmied. Cars may be stolen in spite of them. But no thief ever attempted to steal a car with a man at the wheel. [The Rubber Driver] is so lifelike and terrifying, that nobody a foot away can tell it isn't a real, live man. When not in use, this marvelous device is simply deflated and put under the seat."[40]

Owners were advised to lock their doors or "garage" their automobiles. In his 1917 article "Automobile Thefts," John Brennan proposed another countermeasure: "If owners would only take steps to put private identification marks on their cars, the problem of automobile thievery would be a simple one to solve."[41] It was suggested that owners bore holes in the underside of the running boards, scratch their name in a secret place, or tape an identification card inside the upholstery.[42]

In addition to leaving a secret mark or set of identification marks, stronger locking mechanisms were proposed. One such deterrent, first marketed in 1920, was the Simplex Theftproof Auto Lock.[43] Advertised with the slogan "To be simple is to be great," the device appeared to be a simple collar lock made of bronze and steel, which could be installed in twenty minutes by a mechanic. Once in place, the Simplex lock positioned a vehicle's front wheels straight ahead, and it was claimed

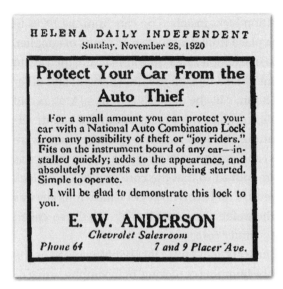

HELENA DAILY INDEPENDENT
Sunday, November 28, 1920

Protect Your Car From the
Auto Thief

For a small amount you can protect your car with a National Auto Combination Lock from any possibility of theft or "joy riders." Fits on the instrument board of any car—installed quickly; adds to the appearance, and absolutely prevents car from being started. Simple to operate.

I will be glad to demonstrate this lock to you.

E. W. ANDERSON
Chevrolet Salesroom
Phone 64   *7 and 9 Placer Ave.*

There were an incredible number of locks of various designs employed during the early 1920s to stop auto thieves.

that such a vehicle could not be towed. Available in five diameters, the antitheft collar could be attached to virtually any American car at the modest cost of fifteen dollars, not including professional installation.

A far more effective and more popular locking device marketed and installed during the 1920s was the Hershey Coincidental Lock. In a 1928 advertisement in the *Saturday Evening Post,* its manufacturer asked, "Will your NEW car be safe?" The ad copy argued that the product— with more than 2 million sold—locked "not only the ignition, but the steering as well—with a hardened steel bolt." The Hershey Coincidental Lock was the work of inventors Orville S. Hershey and Ernest J. Van Sickel, both from Chicago.[44] First developed in the early 1920s and then refined during the remainder of the decade, the lock anticipated steering wheel locks that were mandated by the federal government in the 1970s; it was perhaps stronger than the locks that appeared on Big Three cars at that later date. With a combination ignition cutoff and a strongly reinforced deadbolt, the owner of a Hershey automobile lock could opt to disable the deadbolt and secure the car only by switching off the ignition or could employ both deterrents.

One might hastily conclude that manufacturers had little interest in selling automobiles equipped with secure locking systems, but that would be wrong. It is difficult to generalize on the matter of original-equipment ignition locks installed on cars from the 1920s to the 1930s. Changes took place from year to year in terms of supplier, design, and placement. For example, for a time in 1932 and then again later, V-8 Fords sometimes had a lock on the transmission and other times on the steering column.[45] By the mid-1930s General Motors had settled on a disk or wafer tumbler design, based on the one developed by Briggs and Stratton. This lock featured six wafers (pins were also used in some cases) and a single-sided key. If the proper key was inserted, the wafers were moved so that the core or plug of the lock could freely rotate. However, if there was no key or if the wrong key was placed in the cylinder, the wafers were not aligned with the so-called lock shear line, and the cylinder core remained fixed.[46]

But, no matter how intricate, locks were invariably defeated by the experienced thief, either by bumping or picking, or by simply cutting or forcing them open. Another approach tried in the 1920s was to provide a visible identification number and a set of authorized driver photographs. Such was the product marketed by the Auto-Thief-Stopper Company of Detroit. The invention of Wallace C. D. Cochran in 1922, the "STOP THE THIEF" plate was secured to the vehicle's gas tank. The information on the specially embossed and sealed card included a "Whizzer" serial number, photographs of owners and authorized drivers, their addresses, and information on the color of their hair and eyes, complexion, and distinctive marks that might include birthmarks and scars. The basic notion was that service station attendants would check the plate before servicing the car and "if any doubt [arose], hold parties and summon an officer."[47] If anyone attempted to remove the plate, the result would be a hole in the gas tank; altering the plate would break seals that could not be repaired. In the historical record there is no evidence that such an identification card ever caught on during the 1920s. But the plate did reflect one of the most important shortcomings of the automobile of the 1920s: that a uniform system of vehicle identification numbers did not exist and that numbers stamped on the motor or chassis were easily altered.

Perhaps the most significant antitheft technological system introduced during the 1920s aimed at owners and manufacturers was the FEDCO number plate. According to one company brochure, the

FEDCO system was a response to the utter failure of any method to arrest the alarming increase in auto thefts during the mid-1920s. The New York City–based firm FEDCO (Federated Engineers Development Corporation, also known as the Fedco Number Plate Corporation) was "devoted to the complete, practical development of inventions." Beginning in 1923, it had worked on an antitheft number plate with the Society of Automotive Engineers, the Underwriters Laboratories, and the Burns International Detective Agency. Essentially, FEDCO metallurgists fabricated a plate that self-destructed when one attempted to remove or alter it. With digits made of white metal that stood out from a background of oxidized copper, the entire plate also had an embossed surface that was characteristic of the car make. Below the numbers the digits were spelled out. This complex identifier perplexed those who tried to foil it. The idea was to attach a number plate to the dashboard of a new car coming off the assembly line. To alter or remove the plate would result in its destruction, leaving a telltale remnant indicative of tampering. Apparently the technology worked, at least according to one car thief who was caught as a result of it. Writing his confession from the Nassau County, New York, jail, auto thief A. M. Bachmeyer exclaimed, "Had I realized just what this number plate meant I would not have stolen this car."[48] While the FEDCO system proved to be an effective deterrent, there is no evidence that Chrysler continued with the number plate after 1926 or that other manufacturers adopted the unique plate technology.[49]

THE DYER ACT (1919)

The automobile-caused increase in mobility was matched by growth in the size of government to counter automobile theft. "To cope with this problem," Arch Mandel wrote in 1924, "police departments have been obliged to detail special squads and to establish special bureaus for recovering stolen automobiles. . . . This has added to the cost of operating police departments." The cost of police work in cities with populations over thirty thousand rose steadily from approximately $38 million in 1903 to $184.5 million in 1927. The cost of police work in state governments increased from approximately $98 million in 1915 to $117 million in 1927. To combat auto theft, state governments created license, registration, title, and statistical bureaus and urged the federal government to become involved. E. Austin Baughman, Maryland commissioner of motor vehicles, cited 1919 as "the climax of an epidemic of car stealing," with 922 cars stolen, 709 recovered, and 213 missing. Baughman urged the country to adopt a title law, and by 1919 the states of Michigan, Indiana, Virginia, Delaware, Missouri, North Carolina, Florida, Pennsylvania, and Maryland had such laws on the books. The title laws made it possible to identify and locate all motor vehicles through the name and address of the owner on record.[50] In 1920 Massachusetts developed a similar program under the used-car department of the Department of Public Works. States that did not pass title laws were a nationwide liability and were called "dumping grounds" by neighboring states.[51]

Given the losses incurred by insurance companies and the interstate nature of automobile theft, federal intervention was forthcoming. As Arch Mandel wrote in 1924, "State lines have been eliminated by the automobile [and the] detection of criminals is becoming more and more a nation-wide task."[52] In 1919 Congress passed the National Motor Vehicle Theft Act, which received the appellation of its sponsor, Senator Leonidas Dyer. In a speech promoting his bill to the Congress, Dyer asserted: "It is getting so now that it is difficult for the owners of cheaper cars to obtain theft insurance due to the great loss that insurance companies have sustained. During the past year, automobile theft insurance on this class of cars has increased 100 percent."[53]

Yet, if we are to believe one source, it is not Dyer but the insurance industry that should be credited with this important piece of federal legislation.[54] In essence, the motor vehicle greatly extended federal crime-fighting power across state lines. In 1918 Chicago insurance investigator E. L. Rickards visited Michael Doyle of the American Automobile Insurance Company, and it was Doyle's connection with Dyer that resulted in a law that was a response to the problem of the interstate transportation of stolen vehicles. Enforcement of the Dyer Act became the responsibility of the U.S. attorney general and the Bureau of Investigation, although it took several years and some prodding on the part of insurance interests before federal authorities became actively engaged in its enforcement.[55] Details about jurisdiction and venue remained to be resolved. One question that quickly surfaced was where an automobile thief would be prosecuted: at the place where the car was stolen or at the point of apprehension. NATB officials went to Washington and met for two days with Bureau of Investigation director William J. Burns, Bureau of Investigation assistant director J. Edgar Hoover, and Assistant Attorney General Jess Smith. As a result of the conference, it was decided that the thief would be prosecuted at the point of apprehension. Furthermore, the Dyer Act promulgated that thieves would be fined five thousand dollars or sentenced to ten years in prison, or both. The American Automobile Association and the insurance industry lobbied Congress to pass the Dyer Act, which proved to be key to federal policing in all kinds of crime, especially bootlegging and bank robberies during the 1920s and the early years of the Great Depression.[56]

Controversy dogged the Dyer Act from its beginning to the post–World War II era. One main point of contention in the courts and on the part of many an accused auto thief centered on the phrase "knowing to be stolen," in sections 3 and 4 of the law.[57] Did the word "stolen" in the statute also apply to embezzlement and common-law larceny? Local, state, and federal courts and the Department of Justice were asked to interpret the intent of Congress in numerous cases involving both joyriders and organized rings. While the phrase "knowing the same to be stolen" was sufficient to include the act of joyriding, post–World War II legislators, concerned over juvenile delinquency and the rise of white, middle-class crime among youth, discussed narrowing the act to restrict the offense to more serious crimes. Hundreds of boxes of correspondence at the National Archives contain documents that reflect the ambiguity of the law and the subtle nuances involving everyday people.[58] Cases of misunderstandings concerning the use of a car, the purchase of vehicles not known to be stolen, and harsh justice directed toward juveniles were frequently and painfully recounted in letters to the attorney general, J. Edgar Hoover, and President Franklin Roosevelt. On local and state levels, the wording of legislation was often rephrased so as to create a new offense, "the operation of a car without consent of the owner."[59] Whatever the motives and legal wrangling, auto thefts were the most prominent federal prosecutions of interstate commerce between 1922 and 1933.[60]

In 1926 Henry Chamberlin, president of the American Institute of Criminal Law and Criminology, reported a considerable increase in legislative activity in various states in the face of the ever-increasing incidence of car theft. The heightened legislative activity at the state level reflected the beginnings of a modification in the general perception of the identity of the car thief. Up to this time, there was fairly general agreement among experts and the public that the car thief was a professional criminal who worked largely as part of a gang—a predator who would be deterred by the threat of severe punishment. By the late 1920s, however, some correctional administrators and legislators began to question that image of the car thief. In a few states such questioning was reflected in legislation that recognized the difference between an offender bent on financial gain and a youngster bent on pleasure.

HOT CARS

Thieves stole a range of models, mostly low-priced Chevrolets, Plymouths, Chryslers, and Fords.[61] Table 1.1 lists the makes and numbers of cars that were reported stolen in Buffalo, New York, between May 15 and July 15, 1924.[62] The Automobile Protective and Information Bureau reported a similar distribution of makes and models stolen during 1920–21 on a national scale; recovery figures averaged less than one-third of the vehicles initially taken. Fords were the overwhelming choice of car thieves, followed by Buicks, Hudsons, Cadillacs, Chandlers, Studebakers, and Overlands.[63]

Place mattered. Automobiles were most likely to be stolen in business or entertainment districts, where individuals would park the same model in the same place. Often, thieves caught red-handed simply claimed that they had hopped into the wrong car. When interrogated by a judge, one alleged thief explained why he was in the wrong Ford: "Because both cars are Fords, and all Fords look alike, not only to me

A stolen Hudson motor car, about 1920, recovered after being left in the middle of an intersection. Courtesy Library of Congress.

but to their owners."[64] Charges were dropped. Despite preferences to steal the commonplace vehicle, elite and unusual automobiles such as the Auburn, the Cadillac, the Franklin, and the Packard were not exempt from the threat of theft. Expensive cars were stolen, disassembled, and repainted.

JOYRIDERS AND PROFESSIONALS

Quite different in terms of criminal intent were the activities of the so-called joyrider. Joyriders stole for thrills. In 1917 George A. Walters, secretary to the Detroit chief of police, estimated that 90 percent of Detroit's auto thefts were performed by joyriders.[65] Joyriders were often groups of young men in pursuit of fun who had a "taste for motoring."[66] One author argued that joyriders (in all cases male) had a sexual motivation: "Some young fellow with sporty tendencies and a slim pocketbook wants to make a hit with some charming member of the opposite sex. . . . He thinks an automobile would help him in the pursuit of her affections."[67] After a joyride, automobiles were often found damaged and out of gas. Historian David Wolcott noted that in Los Angeles, "boys approached auto theft with a surprisingly casual attitude—they often just took vehicles that they found unattended, drove them around for an evening and abandoned them when they were done—but the LAPD treated auto theft very seriously."[68] Joyriding, usually the act of a delinquent, was so serious that young boys were prosecuted under the Dyer Act of 1919. The federal government did not draw a distinction between joyriding and professional auto theft until 1930. Congressmen Dyer called for the repeal of his own law, and to persuade the U.S. House of Representatives of the need for repeal, he read a letter from the superintendent of a penitentiary: "Of the 450 Federal Boys in the National Training School here in Washington, nearly 200 are violators of the Dyer Act, with the ages distributed as follows: Two boys 12 years of age, 6 boys 13 years of age, 19 boys 14 years of age, 31 boys of 15 years of age, 64 boys 16 years of age, 48 boys 17 years of age, 19 boys of 18 years of age, 1 boy 19 years of age, and 1 boy 22 years of age."[69]

Owing to the capricious nature of theft for a joyride, policemen and journalists surmised that it could be easily prevented: "It is against this class of thief that the various types of automobile-locking devices and hidden puzzles are effective. . . . Since the joy rider does more than half the stealing it follows that car-locks are more than 50 percent effective in protecting a car."[70]

However, more elaborate means were necessary to stop the professional thief. Apparently, the joyriding problem temporarily declined during the 1930s, but organized gangs emerged as a more serious threat, vexing authorities. Car-theft gangs or rings could be found in virtually every American urban center during the 1920s and 1930s.

University of Chicago social scientist John Landesco spent five years during the late 1920s and the early 1930s studying organized crime, including auto theft.[71] He focused on two generations of youth who were a part of a particular neighborhood, the ethnic Italian 42 Gang. One of its members described his experiences in considerable detail. Landesco's work traced the evolutionary nature of criminal behavior as he studied the development of these individuals from small boys who knew each other as playmates to young adults working various rackets. Young men in the 42 Gang started their criminal activities by stealing silk shirts from clotheslines in the western suburbs; only later did they graduate to stealing cars for bootleggers. In between they shot dice, broke into vending machines ("peanut machines"), and stole bicycles, car and truck tires, and butter and egg trucks. Gang members quickly learned that if caught they were to keep quiet, endure beatings, and buy off the cops. Rocco Mercantonio, a prominent member of the 42 Gang, recounted that his auto-theft activities involved stealing cars to order and getting paid "$75 to $200 per car—Fords $75, Buick or Chrysler for $150, a Peerless and Packard for $200."[72]

Writers who addressed auto theft from 1915 to 1938 admitted that professional thieves could not be stopped. They employed an array of tactics to steal automobiles. Often chauffeurs, mechanics, and garage men became thieves. Locks supposedly prevented theft by joyriders, but professional thieves would simply cut padlocks and chains with bolt cutters. In a May 1929 article entitled "Tricks of the Auto Thief," *Popular Mechanics* described an array of tactics. Thieves unlocked and started cars with duplicate keys, "jumped" the ignition by placing a wire across the distributor and coil to the spark plugs, and towed cars away. The article explained, "Some thieves make a specialty of buying wrecked or burned cars as junk. . . . They receive a bill of sale, salvage parts which they place on stolen cars, and so disguise the finished automobile as a legitimate car for which they have the bill of sale." In one method, "kissing them away," an individual would break into a car, and when he was unable to start it, a "confederate" would push the stolen car with his own from behind. The car would be moved into a garage or an alley and promptly dismantled. Thieves used interchangeable parts to confuse authorities. In 1925 Joe Newell, head of the Automobile Theft Bureau in Des Moines, Iowa, stated, "The greatest transformation that takes place in the stolen machine is in the clever doctoring of motor serial numbers. . . . This is the first thing a thief does to a car." Automobiles were branded with a serial number that corresponded to a factory record, a number that preceded what is now known as the Vehicle Identification Number (VIN). However, thieves used several tactics to change the numbers. Numbers could be "doctored" by filing them down and branding a new number into the car or by changing single digits. William J. Davis explained, in a detailed article entitled "Stolen Automobile Investigations," "It is possible for a thief to re-stamp a 4 over a 1; an 8 over a 3 where the 3 is a round top 3; a 5 over a 3; to change a 6 to an 8, or a 9 to an 8, or an 0 to an 8." In *Popular Science Monthly,* Edwin Teale stated: "The automobile stealing racket in the United States has mounted to a $50,000,000-a-year business. During the first six months of 1932, 36,000 machines disappeared in seventy-two American cities alone. In New York City, $2,000,000 worth of cars was reported stolen in 1931."[73]

Over time, professional auto thieves came up with their own lexicon. A "Bent" or a "Kinky" was a stolen car; a "Breezer," an open tourer; a "Clean One," a car from which all identifying numbers were erased; a "Consent Job," an automobile stolen with the owner's consent for the purpose of collecting insurance; a "Dauber," a car painter who worked quickly; a "Dog-House," a small garage used as a safe house for stolen cars until they cooled; a "Right Guy," a dealer who buys stolen cars; a "Slicker," a stolen car newly painted; and a "Stranger," a stolen car taken from a distant location. Auto thieves also purportedly had their own proverb, "Never steal anything you can't steal right away."[74]

Gangs developed sophisticated automobile-theft operations, from the expert driver to the expert mechanic.[75] The "Clouter" actually stole the car, and the "Wheeler" drove it to the "Dog-House." The thieves were concerned with stealing the popular, mid-priced, widely used makes. Gangs often specialized in a certain make or model. One New York gang "Scrambled" the stolen automobiles: "A number of machines of the same make and model are stolen at the same time. . . . Wheels are switched, transmissions shifted, bodies changed, and engines transferred from one car to another."[76] Other times gangs used the "Mother System," in which thieves stole a certain make, had a fake bill of sale made, and changed all of the serial numbers to be identical with the bill of sale. Ultimately, four or five of the same car existed, all with the same serial numbers and bills of sale.[77]

Rings existed in virtually every major American city during the interwar years, and several of the more sophisticated organizations had connections in multiple major urban areas as well as business links to affluent suburbs. With the coming of the Great Depression, desperation caused even the respectable to participate in gangs, often in leadership roles. For example, William Stanley Hayes, a forty-four-year-old Yale graduate, was arrested in 1930 for acting as a "high-powered" salesmen for a gang active in New York, Chicago, and the Connecticut suburbs. Hayes, who claimed to work for two persons higher up in gang hierarchy whom he could not positively identify, joined the ring after his work as a jewelry salesman had dried up and he found himself several months behind with his hotel rent. Uncovering prospects, demonstrating cars that were already in stock, or arranging for the stealing of cars to order, Hayes offered stolen vehicles at attractive prices—some 40 percent off list value. With altered license plates and motor numbers, these cars were often difficult to trace. It was only by chance that Hayes was caught. He was unlucky enough to be walking down the street when a duped customer, in the company of police, ran into him.[78]

Professional auto-theft rings often offered to supply stolen cars to order for customers. One well-equipped gang of three in Brooklyn, for example, used its supply of bogus bills of sale, motor vehicle department forms, and an ingenious "jumper box" to steal and sell some twenty-five cars in only two or three months. Guaranteeing delivery within twenty-four hours of order, the trio included a Brooklyn Sanitation Department worker who served as "spotter" while the other two used an assortment of tools to pry open locks and connect starter motors. The jumper box, which police described as being unique, featured a magneto and wires to connect a vehicle's distributor with its starter, thus bypassing the ignition switch.[79]

The more sophisticated the gang, the more sophisticated the technology employed and the broader its geographical sweep.[80] One New York City gang stole most of its cars in Connecticut and then brought the vehicles to Manhattan.[81] Another ring took vehicles from the suburbs and disassembled them in a plant in New York City.[82] If auto parts were not readily disposable or were too bulky to process, they were dumped on what became the site of the 1939 New York World's Fair.[83] In both of those cases, the rackets were family affairs, employing husbands, wives, and brothers, whose day jobs were car salesmen, chauffeurs,

The cover of a pamphlet published by the insurance industry in 1921. Courtesy National Insurance Crime Bureau.

A GROWING CRIME!

and garage owners. And there was the New York City–based gang that operated in some adjacent states and possessed, among other things, notary seals and a stamping machine. The *New York Times* reported that this group of car thieves and dishonest dealers had profited to the extent of about $750,000 in a complex operation involving precisely counterfeited motor and serial numbers as well as the paper and ink for registrations and bills of sale.[84]

The American mafia certainly had a presence in auto-theft activities before World War II, but FBI leadership continued to deny the existence of a widespread, tightly-knit organization of criminals until the late 1950s. One author has asserted: "J. Edgar Hoover and the FBI on one hand, and the Mafia on the other, grew and prospered together, neither causing the other the slightest anguish. This has been the greatest, most obvious and most inexplicable failure of the FBI."[85] The success of the mafia was its ability to gain political protection for its activities, and until that political protection broke down in the late 1950s, it was immune to prosecution and the curtailment of its various illegal activities.[86]

Hoover's primary target in the years both before and after World War II was undoubtedly in the mafia circle. In 1936 Hoover penned an article about the gangster and international car thief Gabriel Vigorito (a.k.a. Bla-Bla Blackman). Blackman had amassed a $1 million fortune from automobile theft.[87] His Brooklyn-based business marketed his hot cars to Persia, Russia, Germany, Norway, Denmark, Belgium, and China. In 1934 authorities sentenced Bla-Bla to ten years in prison, but that did not end his career; it extended to the mid-1950s. Historically, the point is poignant; the automobile trumped not only state lines but also national boundaries.

The rise of the global automotive industry paralleled the growth of global theft rings. In 1936 the Roosevelt administration made a treaty with Mexico for "the recovery and return of stolen or embezzled motor vehicles, trailers, airplanes or the component parts of any of them."[88] The treaty prompted a convention with Mexico in 1937 to address the stolen-automobile problem, which went back to the time of the Mexican Revolution and intensified after World War II.[89]

FIGHTING BACK

To control rampant automobile crimes, authorities developed scientific means to fight back. As early as 1919, a system of fingerprints to identify automobile owners was proposed. In 1924 the FBI constructed a vast (for the day) fingerprint database called the National Division of Identification and Information.[90] Starting with more than eight hundred thousand fingerprint records, the database grew beyond 2 million by the end of the 1920s. With fingerprints filed, classified, and categorized, the next step taken was the compilation of crime statistics. In 1930 a system of uniform crime reporting began. One of the offenses included in what became the monthly Uniform Crime Reports was auto theft. The result was a comprehensive but imperfect picture of the amount and types of crime committed in towns and cities across the country. Other measures followed.

Stolen cars were central to FBI activities well beyond tracking down the common car thief. For example, when John Dillinger broke out of the Crown Point, Indiana, jail, it was his theft of a car and crossing of a state line that brought agent Melvin Purvis into the Dillinger hunt. And a stolen car matter was often the first case on which a new FBI agent cut his teeth. Purvis, who began his FBI career in Dallas, later recalled in his memoir, *American Agent,* how by glancing at phone numbers scribbled on a restaurant wall and with some luck and cunning, he tracked down his first criminal, an elusive auto thief. It was an episode that augured well for his future as a G-Man.[91]

Between 1920 and 1941 police departments developed increasingly sophisticated means to monitor a more mobile public. Routine methods employed in Buffalo, New York, during the mid-1920s included the use of police radio, as well as daily and weekly "hot sheets."[92] In a similar fashion, Los Angeles police department officers began their shift with a list of stolen automobiles printed the night before. In 1927, a Toledo, Ohio, officer suggested to his superior a system that was remarkably close to the one ultimately adopted by the Auto Theft Committee of the International Association of Chiefs of Police:

1. Get full report and Description of auto stolen. As to date and time when auto was reported stolen. Get correct Motor and serial numbers. And license numbers.

2. After the auto has been out for a period of 48 hours, have Insurance Company put the alarm card out.

3. If not insured, have the owner of the auto order through the Police Dept. where auto was stolen from get out about 250 cards. These to be sent to all Dept. that keep a stolen auto file on autos.

4. These alarm cards should be of regulation size, size 3×5. The card for mailing can be a large card, but should be printed so they can be cut to 3×5. And the Motor number should be in the upper right hand corner. Where it can be seen at once. This will help to speed up the filing and checking of autos. And the name of the make of the auto should be close to the top of the card.

5. The officers should check more autos that they see around town, mostly the autos that have out of town license plates they should

check these parties to their bill of sales, or the license slips that they are given by the state.

6. Another way to check stolen autos would be to lift the hoods of autos parked on the streets at night, also could be done during the day, mostly around the larger factories. We located some autos by this method and have apprehended some good auto thieves. Men that have made a business of selling autos with changed Motor numbers, to some men that they are working with.

7. Every bill of sale and Title form from another State, should be checked to see if the Party filing this auto in the State, has the right auto in his Possession. It is very easy nowadays to buy a junk auto, and get the title and then steal another auto, and go to some other State and Register this auto. We have been doing this for one year now, and find it is quite a success.

8. Another way to check on stolen autos is that when a party applies for License plates, to have these autos checked by police department and see that everyone [who] gets a license number must have the auto in his or her possession. With Proper papers. I understand they are doing this at Los Angeles Cal and they have quite a success in get[ting] back stolen autos.

9. Every Police Dept. should send out a list of every auto that has been recovered. About once a month. As to keep the files clean of autos that have been recovered.[93]

Police departments also developed processes using chemicals and torches to identify fake serial numbers.[94] These destructive forensic techniques enabled investigators to recover filed or ground-down motor numbers that were seemingly hidden to the naked eye. Typically, the stamping process involved striking a cast iron motor block with a die that deformed the metal, leaving an impression. Physically, two zones of deformation resulted, one called the plastic, and another deeper area, the elastic. Even if the number was totally removed from the plastic layer, it still existed in the elastic, and techniques were developed to make those digits visible again. One popular technique used etching solutions, most commonly Fry's Reagent (90 g copper chloride, 120 ml hydrochloric acid, and 100 ml water), which when swabbed on the surface of the damaged metal revealed the erased numbers under reflected light. On a horizontal surface such as an engine block, a frame of modeling clay was made around the area in question, and into it 2 to 5 mm of etching solution were poured. A heat-treatment method was also quite successful in recovering filed-off numbers on cast iron substrates. Using a propane torch, a technician heated the metal to cherry red, which caused the deformed area to bulge above its surroundings. Marks were thus pushed up, and after cleaning, the numbers stood out in contrast to darker surroundings.

These developments were later supplemented by enhanced communication technology. In 1936 it was urged that "every city join the nation-wide network of inter-city radio-telegraph service provided for by the Federal Communications Commission."[95] *Popular Science Monthly* advised: "Chattering teletype machines and short-wave radio messages outdistance the fleetest car, while police encircle a fleeing criminal in an effort to make escape impossible." Without doubt, radio communication made auto theft difficult, but some thieves devised strategies to overcome police radios, including using radios themselves. By 1934, according to one overly optimistic and naive journalist, "auto thieves found their racket a losing one." In response to mobile crime, governments at all levels grew more sophisticated. Insurance companies responded in kind: "In Chicago, a central salvage bureau, maintained by insurance companies [was] established in an effort to wipe out a $10,000,000-a-year racket in stolen parts."[96] Automobile manufacturers invested in a "pick-proof" lock.[97] From 1933 to 1936, insurance companies and the government decimated the market for stolen automobiles and parts. *Popular Science Monthly* reported in 1934: "Figures compiled by the National Automobile Underwriters Association show that eighty-six percent of the cars stolen in 1930 were recovered while in 1931 82 percent were recovered and 89 percent in 1932."[98]

As shown in table 1.2, there was a sustained decline throughout the 1930s in urban thefts known to police in cities with populations over twenty-five thousand, suggesting that organization, technology, and perhaps depressed economic conditions were contributing to what amounted to good news. Along with the end of Prohibition, this optimistic turn was possibly the reason why depictions of stealing cars in Hollywood films released before World War II tended to play incidental roles in comic or farcical situations. In films including *New Adventures of Get Rich Quick Wallingford* (1931), *It Happened One Night* (1934), *Social Error* (1935), and *Bringing Up Baby* (1938), the serious side of automobile theft, more characteristic of the tone of the 1920s, was dismissed.

Yet the confidence exuded by government crime experts during the 1930s was short-lived. In sum, up to 1941 the combination of technology and organization was able to abate, but not stop, thefts of something Americans had learned to love. The problem of auto theft in the United States surged after World War II with a vengeance. For a time, the issue centered on youth, first those from the middle class and then from minorities. Later, professional criminals were a target of law enforcement, but in a cultural climate that increasingly portrayed the auto thief as a hero. After all, the automobile in America was the ultimate freedom machine, and driving proved as exhilarating and liberating for thieves as for owners.

With a shortage of rubber and rationing during World War II, it was no surprise to discover this abandoned car with tires stolen in 1942. Courtesy Library of Congress.

2

Juvenile Delinquents, Hardened Criminals, and Some Ineffectual Technological Solutions (1941–1980)

As a first step in effecting a more concentrated offensive against automobile thieves, the FBI called upon state and local law enforcement agencies to meet with its agents in regional conferences, which are now being held throughout the nation. Also participating in these conferences are state motor vehicle bureaus, the National Automobile Theft Bureau, and other interested agencies. Devoted solely to open forum discussion of car thefts, the conferences are meeting everywhere with interest and enthusiasm. . . . There has been universal agreement that an alert, educated public is the greatest asset available to the law enforcement officer in coping with this type of crime.

J. EDGAR HOOVER, 1952

Regardless of the era, any critical discussion related to the topic of auto theft defies simple explanations. For example, at the beginning of World War II, many social commentators spoke of a general outbreak of juvenile delinquency as a sort of mass hysteria. One might expect that auto theft and joyriding had increased after Pearl Harbor. Nothing was further from the truth, however. Juvenile auto theft decreased markedly during 1942, probably owing to the actions of more careful owners who were mindfully aware that tires and gasoline were in short supply and that insurance companies were refusing to replace either stolen tires or stolen cars. But 1942 proved to be only a lull in the incidence of auto theft, for in 1943 the trend reversed, perhaps a result of wartime prosperity, as more teenagers owned hopped-up cars and less care was given to either driving or parking. Sociologist David Bogen concluded from the experience of the wartime years that "the tendency of delinquency to increase in times of prosperity and decrease in times of economic stress and unemployment is just the opposite of what most people believed."[1] Interestingly, Bogen's wartime observations hit upon one of the most significant paradoxes related to auto theft during the postwar years: that the privileged, white teenagers with money were more likely than their poorer contemporaries to steal cars, and they did it more for youthful adventure than necessity.

The car-theft problem intensified in the years immediately after World War II. In Milwaukee, where in 1946 car thefts were double those of 1945, a group of community leaders that included the chief of police, a Catholic cleric, and a Protestant clergyman formed the Milwaukee Metropolitan Crime Commission.[2] This group was "particularly concerned with the prevention of crime among juveniles" and to that end pursued a public education campaign that resulted in the publication of two brochures, *Here, Kid, Take My Car* and *Keys to Another Kingdom.* Their message was a simple one: remove the ignition keys and much of the juvenile crime will disappear. Thus, the burden was placed on adult automobile owners, who were seen as being even more disrespectful of law and order than youth. In its brochures, the commission firmly argued that "opportunity makes the thief" and that in certain locales up to 92 percent of cars stolen were the result of keys left in the car. *Here, Kid, Take My Car* pointed out: "No citizen . . . would leave a thousand dollar bill lying in the street and expect it to be there when he returned. And yet we have several hundred citizens who daily leave a valuable automobile on the streets . . . with plenty of gasoline in the tank and keys in the car."

The magnitude of the crime was reflected by the fact that in 1946 21.8 percent of all federal prisoners were auto thieves, representing the single largest group of criminals in federal custody. A decade later, the number of prisoners had doubled. Testifying before a Senate subcommittee on juvenile delinquency in 1955, Director of Federal Prisons James V. Bennett remarked, "It is around the automobile, by far and away, that the largest number of federal offenses revolve."[3]

From a broader view, between 1945 and 1980 car-theft trends fluctuated, as many technical, legal, and organizational efforts—some more successful than others—attempted to reduce theft statistics.[4] Prevention and recovery measures became increasingly multifaceted. They included antitheft devices, information systems, insurance industry–sponsored economic incentives, government regulations placed on manufacturers, and criminal laws, as well as police and community programs. Yet all of these efforts served only to slow down lawbreakers, who engaged in a wide range of illegal activities that included joyriding, carjacking, theft for profit, chop shops, salvage-yard fraud, and exportation of stolen vehicles. To focus on car theft and characterize it in the post–World War II era is to aim at a moving target, always changing in terms of location, technologies, methods, perpetrators, and motives.

Nevertheless, by 1970 car crime had shifted from largely youthful joyriding to professional and organized crime. The key statistic in distinguishing amateur from professional activity was the recovery rate. The lower the rate, the more organized criminals were involved; the higher the rate, the more the crime was attributed to the delinquent.

Aggregate numbers of auto thefts in the United States since 1945 show an "increase generally until 1992, after which there is a precipitous decline."[5] In 1948 there were 165.5 auto thefts per 100,000 people.[6] The FBI Uniform Crime Reports indicated that there were 500 car thefts per 100,000 cars in 1960 and a plateau of approximately 1,000 thefts per 100,000 cars between 1968 and 1984.[7] As shown in table 2.1, the total number of car thefts increased from 82,866 in 1950 to the astounding figure of 1,097,189 in 1980. But these numbers only partially tell the story.

It appears that until 1970, most car thieves were teenage joyriders. Consequently, there was a very high rate of recovery of the stolen vehicles. After 1970 there was a change, as suggested in a study by Beverly Lee and Giannina P. Rikoski, who concluded from available data that "the rate of stolen vehicle recoveries dropped from 84 percent to 55 percent in just ten years. At the same time, the value of unrecovered vehicles multiplied by a factor of ten, from $140 million in 1970 to $1.46 billion in 1980." Lee and Rikoski concluded that between 1970 and 1980, "increasing adult involvement, increasing thefts of trucks and commercial vehicles, and declining recovery rates . . . [were] strong indicators that vehicle theft [had] become the province of professional criminals. . . . Vehicle theft [could] be 'big business,' offering relatively high profits at low risk." They noted that some of the avenues for professional car thieves included "reselling stolen vehicles here and abroad . . . salvage switch activities and the chop shop where stolen vehicles are dismantled for their parts."[8]

AUTO THEFT AND THE GOLDEN AGE OF THE AUTOMOBILE

FBI director J. Edgar Hoover reported to *Motor Trend* readers in 1952, "Automobiles now are among the largest items on the nation's ledger of annual losses due to theft."[9] Between 1945 and 1952 more than a million vehicles were stolen. In 1951 alone, an estimated 196,960 cars worth more than $190 million were counted as stolen vehicles. The 1950s were considered the Golden Age of the automobile in America, and they were also golden years for auto thieves.

In confronting the auto-theft problem during the 1950s and 1960s, politicians and local law enforcement officials regarded it as largely a youth or juvenile delinquent problem.[10] Organized crime or rings remained a very real threat to motor vehicle owners. However, it was a threat largely subsumed by concerns about the next generation, the future of American society, and the tendency of young people to defy authority and commit largely victimless crimes. The enlightened response toward young offenders that followed emphasized reduced punishment, understanding young adult psychology and sociology, and above all education. The hard line of strong absolute values and harsh punishment was largely thrown out the window, at least in public pronouncements. Prevailing attitudes of the day suggested that criminal behavior was a social disease, the result of a breakdown of society and values, and that the youthful car thief was to be treated with compassion and understanding. To that end, a 1955 Senate report pointed to the high rates of imprisonment under the Dyer Act and argued that juveniles taking joyrides across state lines were never meant to be covered by that law. Rather, these youngsters were considered to have "misappropriated cars with no intent to steal."[11]

Compassion and understanding did little to arrest the meteoric rise of joyriding and juvenile delinquency. It was one more phenomenon that added to the fears surrounding the early Cold War era.

Government bureaucrats were fixated on statistics (as they still tend to be), regardless of whether the numbers were helpful or misleading. And the statistics they were looking at indicated a dangerous trend. For example, in 1948 young persons under age 17 were responsible for 17% of all car thefts, while four years later, in 1952, some 52% of auto thefts were committed by thieves under age 17. This disconcerting trend continued unabated until the mid-1950s. In 1956, of the 28,035 auto thieves who were arrested, some 39% were 15 or younger, 56% were 16 or younger, and 73% were under age 18. And at the federal level, 55.5% of all juvenile cases brought before the courts involved auto theft. By the mid-1950s, teenagers were committing the majority of larcenies and burglaries, a fact that did not augur well for a free and democratic United States engaged in a life-and-death struggle with global communism. Further analysis of auto thefts showed that they were overwhelmingly perpetrated by young males (the ratio of male to female was 120:1) and that the urban rate for such crimes was three to four times as high as the rural rate.[12]

Sociologists William W. Wattenberg and James Balistrieri fleshed out the motives and class origins of these young criminals in an important study published in 1952. Their investigation of violators in Detroit was most disturbing to those involved in law enforcement, the courts, and social work and, above all, the parents of adolescents.[13] By examining the arrest data of nearly four thousand young people in Detroit in 1948, Wattenberg and Balistrieri concluded that the majority of auto thieves in this group came from above-average homes, had grown up in racially homogeneous areas, lived in an economically and sociologically stable home environment, and, perhaps surprisingly, were on the whole better adjusted socially than their peers. Why did they steal cars? As one New York City policeman surmised a bit later in the decade, "for the sheer hell of it."[14] Wattenberg and Balistrieri concluded that the failure of community controls coupled with opportunity, amusement, and trifling punishment were at the heart of this major social problem.

Another study from the period, however, that of Logan A. Hidy, demonstrated the complexities associated with any analysis of joyriding.[15] As part of a survey of boys committed to the Boys' Industrial School in Lancaster, Ohio, for the offense of auto theft, Hidy explored the motives of a group of 98 young men. While some 42 percent stated that they stole cars for fun, a surprising 48 percent claimed that they took a vehicle because they needed temporary transportation—understandably so, since half of that second cohort were runaways.[16] As it turned out, almost all of the stolen vehicles were unlocked and open, and 80 percent had the key in the ignition. Only four of the 98 boys intended to keep the car they had taken, and only one wanted to sell the vehicle.

While the Hidy study went largely unnoticed, the Wattenberg and Balistrieri article became the starting point for many follow-up investigations conducted during the 1950s and 1960s. For example, between January 1952 and December 1954, Erwin Schepses conducted a careful study of 81 boys who had previously been involved in one or more car thefts and were committed to the New York State Training School for Boys in Warwick, New York. This group of younger boys, who had previously lived either in the New York City area or in rural Orange County, New York, were categorized in one of two subsets: one cohort only stole cars; the other stole cars and also had committed other antisocial acts such as larceny, assault, and crimes of a sexual nature. Motives for what was considered mostly impulsive behavior included being influenced by the "Goddess of Speed," inherent restlessness, and an "erotic element." The latter motive was given an interesting twist in a 1960 study by Juvenile Court Judge Albert A. Woldman entitled "Juvenile Thefts and Juvenile Court." He concluded that since "'walking' among young people [had] become a lost art," auto theft was "almost exclusively a juvenile offense," adding that "girls who require[d] boys dating them to have cars [were] responsible for many thefts."[17]

In particular, members of the group considered "pure," those who only stole cars, were decidedly white rather than African American or Puerto Rican. They also had higher IQs and a slightly lower rate of recidivism. Additionally, and again confirming Wattenberg and Balistrieri's findings, the families of the "pure" car thieves were typically economically secure and stable.[18]

While the Wattenberg and Balistrieri study pointed a finger at the white middle class, late-1950s crime data on large northern cities suggested that most auto thefts were due to "The Problem of Negro Crime."

Yet it was not a widely discussed issue; a Chicago judge claimed in 1958 that the silence represented a "conspiracy of concealment." According to this view, the NAACP, along with politicians eager to garner the black vote, conveniently ignored the facts. For example, in Chicago (15% African American) twice as many blacks were arrested as whites; in Los Angeles (13% African American) 48 percent of all persons arrested for major offenses (including auto theft) were black; and in Detroit (25% African American), 66 percent of all those held in the Wayne County jail were black.[19] Los Angeles chief of police William Parker did not mince words when he stated that the "Negro Community" was "his No. 1 crime problem."[20] Furthermore, a widely held assumption was that the underprivileged and minorities were more likely to become hardened criminals.

FICTION OR REALITY?

The joyrider, juvenile-delinquent phenomenon of the 1950s and 1960s was reflected in many cultural expressions. Mirroring a major concern of the day, car theft as connected to juvenile delinquency emerged as a key theme in film. The best of these films also explored the latent anxieties concerning reckless youth. The growing reality of young Americans with access to cars and money presented a danger to the symbolism of mobility and consumption as a legitimate expression of independence. Theft of cars by juvenile delinquents not only threatened hallowed notions of property but also called into question the foundations of a society in competition with global communism. The possession of an automobile by an unfit subject—exemplified by the youngster's act of theft—challenged the notion of driving as an expression of healthy self-determination.

In addition, there was a hint that these "prankish" acts were spurred by a legitimate sense of disappointment with postwar society, a society in which parents were somehow lacking in character and moral values. In short, perhaps it was adult society that was delinquent in its duties, and driving the automobile was in reality a shallow fantasy when it came to identity and independence.

Delinquency, auto theft, the failures of weak parents, and nearly unbearable anxiety were integral themes in *Rebel without a Cause.* The film's most vivid scene centers on a "race to the edge" involving two of the film's central characters, Buzz Gunderson (Corey Allen) and Jim Stark (James Dean).[21] This iconic film of the 1950s is not about car theft per se, but about lost middle-class youth living anxious and aimless lives. The "chicken" scene ends with Buzz's accidental death, as he inadvertently catches his leather jacket on a door handle and two cars plunge to the ocean below. The two cars turn out to be stolen, feature-less, and nondescript. The 1940s vehicles, possessing little value and certainly no outstanding features in terms of style or color, reflect the attitudes toward auto theft during the period, in that the cars have little meaning beyond reckless youthful impulses and destruction. They are fungible pieces of rolling metal. Yet, while the loss of the cars is inci-dental, this particular tale ended in tragedy, with the cars as an after-thought. Just as the automobiles fell into an abyss, it seems American youth of the day, living largely meaningless lives, were heading toward a similar end. In driving the stolen cars, Buzz and Jim seek to assert and define themselves among their peer group. The entire episode ends horrifically, and Jim is psychologically and socially no better off than before the stunt.

A more shocking and one-dimensional presentation of a teenage car rebel emerged in *Young and Wild,* a 1958 B-film double-billed alongside *Juvenile Jungle.*[22] Republic Pictures's catchphrase for the film—"The Switch Blade and Hot Wire Set!"—left little doubt about the depravity of the juveniles who were at the center of the story. Their hair-raising recklessness and bullying signified the appearance of a new breed of sociopathic children who threatened civilization.

The story begins with the theft of a 1957 Ford that has the keys left in the ignition. Thus, an owner's carelessness contributes to a vehicular homicide, since an old woman is run down by that Ford, driven by an unrepentant thug who is aided by two greaser accomplices. In other action, the bullies push off the road a clean-cut young man and his pretty date in a 1957 Ford convertible. Trying to cover up their crimes, the trio, backed up by a crass uncle of one of the youths, lie, intimidate, and threaten not only the couple but parents as well. In the end, the heroine of the story, Valerie Whitman (Carolyn Kearney), decides to take a stand that ultimately leads to the arrest of these three delinquents with violent tendencies.

Since the three were twenty years old, the justice system faced a dilemma as to what to do with them. Rick Braden (Scott Marlowe), "Allie" Allison (Weston Gavin), and "Beejay" Phillips (Tom Gilson) beat up boyfriend Jerry Coltrin (Robert Arthur), assault Valerie, and elude capture. Ironically, perhaps, but not surprisingly, the three seem impotent with young women their own age and cannot attract them. Detective sergeant Fred Janusz (Gene Evans) patiently works the case, deals with reverses, and finally puts the delinquents behind bars. Justice prevails, but only after courage is demanded on the part of innocent citizens.

Perhaps the most realistic portrayal of this form of juvenile delinquency appeared in literature rather than film. Theodore Weesner's 1972 novel *The Car Thief* (chapters of which were published as early as 1967) was a dark and rather disturbing representation of the subject.[23] Weesner's central character is Alex Housman, a sixteen-year-old high school student living in Detroit. Alex is white, sensitive, and of above-average intelligence, the product of a working-class broken home. He lives with an alcoholic father on Chevrolet Avenue, yearns for fellowship with a brother he is separated from, and desires love from school girls, whom he fears. As the novel opens on a day in late October, streets are clogged with slushy snow. Alex is about to steal a 1959 Buick Riviera. "Its upholstery was black, its windshield was tinted a thin color of motor oil." The Buick was his fourteenth stolen car. And contrary to notions that the joyrider is exhilarated, Alex is filled with fear while driving the car:

> the tediousness of driving did not go away. The pressure kept growing until he felt it in his jaws, and he began losing his strength of grip on the steering wheel. His stomach was drawing tighter. It was a pressure, an anguish, which had overtaken him before, but he did not think of that, nor very clearly of anything. He closed his eyes against the feeling and opened them. His jaws felt chilled. He removed his foot from the accelerator, and as the sensation was seizing him, he slammed his palms against the steering wheel,

jarring it, as if a violent striking there might cancel an explosion elsewhere.[24]

Alex takes his cars on rides in the country outside Detroit, hoping he can catch a glimpse of the brother he is separated from in front of a tavern in which the latter now lives with his mother and stepfather. Alternatively, Alex has a desire to lure young girls who attend "country" high schools into taking a ride with him. And it is Eugenia Rodgers's coat left in a 1959 Buick he has stolen that ultimately leads to his arrest at the end of a school day. It's the beginning of a rather horrific set of consequences that first sends him to juvenile detention, then back to his high school, where he is branded as a no-good. Ostracism and a beating follow. Teachers and fellow students are often brutal and rarely understanding or forgiving. As the story winds down, Alex's father, worn out by chronic alcoholism and life in general, commits suicide. Clearly, auto thievery has resulted in nothing but pain and grief, none of the supposed thrills experienced by the middle-class joyrider as depicted in sociological studies. Alex shares none of the stereotypical characteristics of the youthful auto thief of his generation. Weesner's book suggests that in the real world of a blue-collar kid, there is little to celebrate in disregarding the law and much to fear.

Driver education and teenage educational films appearing after World War II also focused on the dangers of joyriding, which was described in the justice-system and insurance-industry literature as a crime of opportunity and mischievousness. But alongside the joyriding auto thief emerged another stock figure of troubled youth: the hot-rodder, motorbike hooligan, and greaser boy. An early version of the troubled teenage joyrider can be seen in the 1940 short *Boy in Court,* which follows a young man through the consequences of his decision to enjoy himself by stealing a car. Similarly, the 1955 short *Teenagers on Trial* reveals what happens when a delinquent youth steals a car and hits the town's beloved police officer. More tragedy follows in the 1956 film *Car Theft,* when three youths spontaneously decide to steal a parked car that has the keys left in the ignition and run from the pursuing police. Educational film impresario Sid Davis's 1961 *Moment of Decision* reprises the same situation. Yet here the viewer listens to the internal thoughts of four young men as they contemplate taking a joyride. What we learn from *Moment of Decision,* and indeed from all the previously mentioned films, is that "to a greater or lesser degree we are all products of our environment. What we do, how we think, and what we feel are largely determined by our past experiences. Roger, John, Paul and Bill are no different than all the rest. We all need social approval and fight against rejection."[25]

The dangers that surround the young are more societal and institutional than personal. As a whole, these films suggest that delinquency results from two interwoven problems. The first is the failure of schools, the court system, and the community to provide sufficient opportunities for constructive endeavors. The second is the consequence of temptations that society presents to children and the inability of overworked or self-interested parents to tend to their children's development. "Both your parents have seemed to work as long as you can remember," the voice-over informs us as Bill is thinking in *Moment of Decision.* "Even so there has never been enough money for the things you wanted." As a fast jazz score conveys the excitement the boys feel, we are left to see the virtually unmanageable psychological dilemma faced by the still-immature youth in a society that emphasizes consumer gratification. The private need for collective approval, symbolized by the peer pressure of the other boys, along with the desire to appear cool and in control to onlookers while driving in public behind the wheel of a fancy convertible, pressures the boys to do what they know to be wrong. In this film and others, the failures of adult society were manifested in the owner's irresponsible action of leaving the keys in the ignition. Only Bill, whose father has taken the time to teach him the virtues of hard work and the benefits of responsibility, is able to resist the temptation.

Subtextual themes of class and masculinity are obliquely alluded to in the 1976 educational short film *Joy Ride.*[26] On the surface, *Joy Ride* replicates the structure and antitheft, juvenile-delinquency themes of earlier educational shorts. The protagonists, Val and Tim, two bored, undermotivated youths, are shown hanging around the local park watching others play baseball. Hunched over and beaten down by frustration, they gripe about how Randy, one of the older, athletic kids playing in the game, owns a car and has lots of girls. Realizing that Randy had left the keys in the ignition of his car, a late 1960s Dodge muscle car, they decide to take it for a spin. Later one of the boys says that the pair "kinda took it." Suddenly the two come alive and gain some confidence, although they still fear calling girls, and argue over which of them has better success. Finally, they invite two young girls to ride with them and drive to a park, where an abandoned car serves as a place to play. As the four enjoy one another's company, they play more like the children they are than the adolescents they are becoming. But on their way to return the borrowed car, Val's speeding attracts the attention of a highway patrolman. Unwilling to stop and ignoring the pleas of the girls to slow down, Val continues to speed as one of the girls struggles for the wheel, and the car runs head-on into the side of a cliff. All of the stock elements are present: social neglect, premature grasping for pleasures of autonomy, and above all insecure masculinity as expressed in the car they steal. Yet also lurking beneath the surface is the suggestion that the handsome, popular, mobile, and manly Randy is more affluent, since these young boys dress in clothes that seem to be more from the racks of the Goodwill Store than from Nordstrom's.

In each of these films, the young joyriders are depicted as wayward, uncontrolled adolescents whose crimes are more dangerous than criminal. Emphasizing the same point, much of the focus of police and insurance-industry efforts to contain joyriding centered on encouraging proper precautions by vehicle owners. In effect, they were highlighting the negligence of adults. The films suggest that the problem of the stolen car was linked to the growing opportunity for young people to indulge in pleasures they are not yet responsible enough to undertake. But they also suggest that facilitating the problem was the failure of adult society to responsibly manage the young and set good examples for them. Yet these films hint at a deeper and more psychologically disturbing possibility for the postwar adult who has accepted the car and automobility as the embodiment of autonomous liberal individuality. The act of owning and driving a car may be a puerile fantasy and the world surrounding it a prison. And one wonders if the owner leaves the keys in the car in response to a subconscious "death wish" to be relieved of the car, or perhaps the owner has a fanciful belief that trust and safety are still a part of the rapidly changing post–World War II nation.

A SNAPSHOT OF THE STOLEN CARS

Table 2.2, based on a Department of Transportation survey done in the mid-1960s, summarizes auto theft immediately before the federal government became involved in forcing the automobile industry to redesign various antitheft features, including steering wheel locks, external door locks, and internal door-lock "buttons"; the industry was also directed to eliminate vent windows.

Clearly, in the 1960s, thieves—primarily joyriders—preferred the common vehicle over luxury and European models. The Chevrolet was the overwhelming choice of car thieves, undoubtedly because of its ignition lock arrangement. The everyday Ford was the next most popular make, followed by the three General Motors marques—Buick, Pontiac, and Oldsmobile. For whatever reason, technology, popularity, or availability, Chrysler products were well down on the list. Thieves also shunned the Ford Falcon, perhaps the first disposable car coming out of Detroit, more than either American Motors vehicles or Volkswagens. As table 2.3 illustrates, older cars—for whatever reason (perhaps because owners cared less about them)—were more likely to be stolen. More than 40 percent of the vehicles stolen in 1967 were six years old or older, more valuable for parts than when sold as a whole car. Again, the preference for older cars may have been due to owner carelessness or lack of concern for devalued property. However, the percentages indicate that late-model cars were objects of desire as well. Hot spots for the "boost" were in front of the owner's home and at a shopping center.

By the 1960s, then, trust and safety in and around one's home were called into question. With more than 13 percent of cars taken from driveways or garages (see table 2.4), thieves posed a very real and immediate threat. Certainly the old practices of leaving the doors open at night and sleeping on the porch during a warm summer night posed an elevated risk that was not present in America before World War II.

The crime of auto theft, in this 1967 snapshot, remained primarily theft by the amateur joyrider; the statistics assembled indicated very little professional involvement. Perhaps the data in table 2.5 should be questioned, since professional criminals did not emerge as a chief concern until a decade later. But note that thefts for the purpose of selling the cars or their parts, as well as thefts directly connected with crimes and escape, were quite low in frequency. It was logical that theft rates increased at night, as shown in table 2.6, since surveillance was far more difficult, there were fewer bystanders, and the deterrents of the built environment could be more easily overcome.

Corroborating evidence from industry spokesmen and insurance investigators indicates that the large majority of thefts took place because the owner left the keys in the car or left the ignition switch in the "on" position, a common occurrence during that era when a keyless ignition-switch position was a feature on all General Motors cars. It was an age when community trust was no longer warranted, given the social dislocations of the decade. And not only did the General Motors ignition lock need no persuasion, but keys were easily obtained, either from mail-order businesses that sold master key sets or from cooperative dealers. As shown in table 2.7, owners played into the hands of would-be car thieves by leaving car keys in unprotected places—in the ignition switch, under mats, over sun visors, or on the car seat. Yet even if owners secured their vehicles, no car was safe, and no owner, no matter what class or race, was immune to the activity of the car thief. If a professional thief wanted a car, that vehicle would be taken; it could easily be towed off to a secured site if necessary (see table 2.8).

For example, in 1964 Senator Barry Goldwater's 1963 Corvette Stingray was stolen at the Dulles airport while he was on a trip to the West Coast. That same year, in several major cities, a publicity campaign with the slogan "Lock Your Car—Take Your Keys" was undertaken with positive short-term results. Both in Boston and in San Francisco, the National Automobile Theft Bureau (NATB) worked with police officials to mount a television, radio, and newspaper effort urging car owners to lock their cars. Parking-meter decals and leaflets included in utility bills spread a message that somewhat arrested the tide of increasing thefts, but the outcome of this endeavor was far from overwhelming.

In spite of the campaign, car theft in Boston increased by 18 percent, while in major cities nationwide the rate was 25 percent during the first quarter of 1964.[27]

THE RISE OF FEDERAL ANTITHEFT LEGISLATION AND THE TECHNOLOGY THAT FOLLOWED

Auto theft during the 1950s and 1960s, then, posed two very different challenges to both the authorities and car owners. One issue was how to stop joyriders and delinquents. For the most part, it was thought that a joyrider could be easily thwarted. If a careful owner were to just lock the car and remove its ignition keys, all but the most motivated of this group would be foiled. Going a bit further, it was also suggested that a car owner secure the most vulnerable design element of a motor car built during that period—the vent window. The catch on the vent window could be secured simply by drilling a hole in the window frame and then driving in a sheet-metal screw. And the owner could install a hidden kill switch as a second layer of defense, in case a thief forced the vent window open.[28]

Discouraging professional auto thieves was an entirely different matter. Organized crime rings brought with them a different set of problems.[29] One strategy employed at the time was to set standards for the location and format of the important Vehicle Identification Number (VIN) plate. Before 1955, a car's VIN could be found in several different places: it might be stamped on the motor, on a door post, on the frame, or on the firewall. It could also be stamped on a plate that was affixed to the vehicle by screws, rivets, or a weld. Since many vehicles had VIN's stamped on the motor, and since many owners changed motors without notifying state authorities, motor numbers were of limited value. That was especially the case between World War II and the mid-1950s, when new and used cars were in short supply. In 1955 a first step was taken to solve the VIN issue. Vehicle identification numbers were placed in the area of the left door hinge; in 1969 the location was changed to the left interior dashboard, where they can be found to this day. By using special rivets, thievery was to a degree deterred.

However, these special original-equipment manufacturer rivets were often found on the black market. As shown in table 2.9, the move to standardize VIN's did nothing in the short run to improve vehicle recovery percentages, since a gradual shift toward professional thievery was taking place at the same time.

One way in which the federal government intervened during the mid-1960s in an attempt to reduce the amount of joyriding was by mandating that manufacturers install technological devices that thwarted the amateur thief. As it turned out, these measures were buried in a wave of federal safety and emissions requirements. What resulted after 1967 were numerous improved designs and new devices that stopped only the most amateurish of joyriders. In the process, manufacturers removed the vent window and column-mounted ignition switches, and annoying key buzzers became standard equipment.

The design of ignition switches that would deter car thieves took a very long time in coming. The only significant new security device during the first decade after World War II was Chrysler's "key operated starter switch," which was introduced in 1949.[30] Manufacturers knew that a high percentage of auto thefts were facilitated by owners leaving their key in the ignition switch, and the manufacturers pointed to that fact as a way to justify their own failure to get serious about making their cars theft-proof. From time to time, door locks were marginally improved, but little else was done. However, in the mid-1960s the Lyndon B. Johnson administration, feeling pressure from angry consumers, wielded more aggressive regulatory power. As a result, beginning with the 1965 model year, General Motors and Ford implemented several improvements in the theft security system on passenger vehicles. Ford redesigned all its locks on 1965 models. The new style contained tumblers at both the top and the bottom of the lock assembly. The lock required a key cut on both edges and kept the car from being unlocked with a jiggler key. General Motors modified Chevrolet and Buick ignition systems to make it impossible for the key to be removed without locking the ignition. The new ignition switch had five positions—accessory, lock, off, on, and start. The key could be removed only when the ignition was switched to the lock position. Furthermore, switch wiring terminals were secured and concealed in a plastic connector that was fixed into position on the back of the switch by three projecting fingers that snapped over the concealed lugs on the ignition housing. This made it more difficult to hotwire or jump the ignition wires and start the car without the key. General Motors also agreed to attach VIN plates using rivets instead of spot welding on all models except Cadillacs. A special rivet, appearing as a distinctive rosette or "rosehead," was used, which made it easier to detect fraudulent or changed VIN plates.[31]

These measures, however, did not satisfy industry critics. The National Traffic and Motor Vehicle Safety Act of 1966 empowered the secretary of transportation to issue safety regulations to be followed by motor vehicle manufacturers. The law was broadly interpreted to include auto theft, since joyriders and transportation users were seen as intimately connected to the problem of auto safety. In a related matter, the Highway Safety Act of 1966 gave the Department of Transportation the authority to issue highway safety standards, and pursuant to this mandate, Standard 19 dealt with motor vehicle titles and thefts. It required a birth-to-death uniform title, so that the title remained with car until it was sold or salvaged. Only when a car was sold or salvaged was the title canceled. The old title was returned to the issuing state, and an inspection was to take place with each titling.[32]

In 1968 Congress finally held manufacturers accountable for auto theft by passing Federal Motor Vehicle Safety Standards 114 and 115. Standard 114 called for manufacturers to equip their products with steering column locks and an ignition key, while Standard 115 mandated that a single VIN marking be placed on the vehicle.[33] Although the steering lock was a highly touted innovation that was intended to greatly dissuade car thieves, in reality it did very little to stop them. The United States was following the example of West Germany, where steering wheel locks were made compulsory in new cars beginning in 1961 and where statistical data tended to indicate their effectiveness. However, in the United States, each manufacturer introduced its own lock design, since Congress did not want to establish a restrictive standard that discouraged technological innovation in the field of theft inhibition.[34] Almost immediately after manufacturers introduced steering locks, car thieves countered by developing low-tech tools such as "slide hammers," or dent pullers, to remove the lock from the steering assembly. A screw was welded to the end of the dent puller, and in one rapid motion the lock was extracted. Alternatively, a screwdriver driven into the lock and turned using vise grips broke the tumblers, rendering the mechanism useless. Ford locks proved to be the most vulnerable; subsequent studies reported that a good thief could circumvent a Ford steering lock in 10 seconds and a lock found in Chrysler products in 30 seconds. The steering column locks in General Motors vehicles took 120 seconds to defeat. If a thief chose the strategy of breaking the lock by twisting, the entire operation took about 5 seconds.

However, the twisting technique only worked on Ford and Chrysler cars, since General Motors and American Motors cars had a side bar that prevented the lock from turning. Car thieves certainly knew the score regarding these steering lock designs, as Ford vehicles with steering locks were twice as likely to be stolen as other brands. A subsequent FBI survey reported that approximately 81 percent of the recovered cars with locks removed or ignitions forced were Fords, while only about 2 percent were Chrysler and 13 percent were General Motors cars. These numbers were so damning that Ford redesigned the ignition steering locks on its 1976 model cars.

QUESTION AND CHALLENGE AUTHORITY

Given the emergence of the federal government during the 1960s as a countervailing force to what appeared to be manufacturers' lax efforts to improve theft deterrents in their vehicles, it seems paradoxical that films of that period push back on the subject with themes that question authority and the law. Unlike the 1950s cinema, early-1960s films neglect the subject of lawless youth and automobile theft. But in 1967 the notion of challenging traditional values burst on the scene with the release of *Bonnie and Clyde*.[35] Along with many other late-1960s films, *Bonnie and Clyde* revitalized Hollywood's use of the automobile as a symbol of autonomous individuality. The film elaborates on earlier themes while also taking the symbolic potential of auto theft in new directions. Centering on the lives and violent deaths of the infamous 1930s bank-robbing duo, *Bonnie and Clyde* inaugurated the transformation of auto theft in film from the subtextual expression of delinquents grasping for autonomy into a gradually more overt use of the subject as a reaffirming act of reclaimed selfhood. Along with other contemporary films that featured personal transportation, such as *Bullitt, The Graduate,* and *Easy Rider,* this story of mobile criminality captured the complex generational response to the bankruptcy of postwar American culture. It did so by recapturing a populist view of the Depression-era past.

Bonnie and Clyde opens with a scene of voyeuristic anticipation. Clyde (Warren Beatty) is hesitantly preparing to steal a car. Bonnie (Faye Dunaway), in the nude, observes from her bedroom window a handsome young man suspiciously lingering around her mother's automobile. Rather than alert her family, she watches curiously, drinking in the scene, anticipating what is to come. It is evident that her curiosity stems from the banality of her own life. Confined in domestic imprisonment, she gazes outside to freedom. To passively consume the thrill of theft, she steps outside to confront the would-be thief. Her purpose, however, is not really to stop Clyde, but to join him. Clyde and the stolen Ford Tudor car are her vehicles to escape the self-destroying oppression of the common, undifferentiated tedium of the everyday world. The act of auto theft is the catalyst of the film, and it is repeated time after time, although it is always a secondary crime to that of bank robbery. Accompanied by the rousing banjo classic "Foggy Bottom

 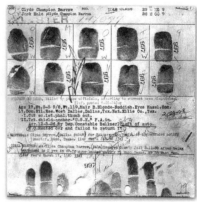

The young Clyde Barrow, first arrested for auto theft in 1926, and his fingerprint chart. Courtesy Federal Bureau of Investigation.

Breakdown," the opening acts of car theft, criminality, and disorderly mobility convey the recapture of control over one's life. What follows is an elaboration of the search to find self through repeated acts of stolen mobility and the defiance of society and its institutions that strangle the chance for personal realization in rural America during the Great Depression.

Set at a time in U.S. history when the failure of society to create an environment in which autonomous individuality might be realized, the film radiates an anti-establishment sensibility. The protagonists become the heroes of the struggling working-class people who live in stark visual settings. As bank robbers, Bonnie and Clyde victimize the institutions that have victimized the common person. In one scene, they are awakened in an abandoned home by the family who has been evicted from it. Hearing their story, Clyde announces to nodding

The case of Karl Clark Strain, 1937–38, whose career was used to illustrate this chart. Strain drove about 22,000 miles within the United States on his crime spree, exemplifying how the automobile made the solving of crimes more difficult. Courtesy Federal Bureau of Investigation.

approval that they are bank robbers and gives the family a chance to shoot the windows of the home now owned by the bank. More deeply, over the course of the film we are led to see that the simple dreams for dignity and independence of these people parallel Bonnie's own basic desires, her wish for a meaningful, loving life. However, the atmosphere of stark emptiness of Depression-era America created by the film makes it clear that Bonnie's wish will not be fulfilled. On the sexual level, Clyde's ability to perform is as disappointing as the landscape. He can shoot a handgun with great accuracy and steal cars effortlessly, but he avoids intimacy with Bonnie at every turn.

Clyde's reasons for stealing cars are rooted in masculine frustrations that mirror Bonnie's simple longings for dignity and fulfillment. Throughout, Clyde betrays a volatile combination of diffidence and rashness expressive of suppressed manhood, a condition embodied in his sexual impotence. Bonnie's assertive, longing sexuality repeatedly spurs him into substitute actions of auto theft, bank robbery, gunplay, and compulsive mobility. His manliness has, in effect, been diverted into bold usurpation and defiance of institutional authority in the form of banks, the police, and the laws of the road. It's their ticket to dignity. Repeatedly, the theft of a car saves them from capture or death by the malignant forces. Thus the theft of automobiles, far more than the money they take, represents their independence. Emphasizing the point is the up-tempo banjo music, signaling the positive restoring vitality of the acts. At the same time, the music's folksiness suggests a populist political message that each act of stealing mobility restores the traditional distinctive individuality that has been lost in mass society.

Yet these themes of restoration of selfhood through stolen mobility are undercut as the film progresses. Like many of the films of the New American Cinema movement, *Bonnie and Clyde* demonstrated a reflexivity that brought into question the conventional use of mobility as a simple signifier of restored liberty. Midway through the film, a darker mood of futility and inevitable doom begins to creep across events, foreshadowing the bloody climax at the end. The gang of idiosyncratic individuals begins to unravel. At various points they are shorn of their vehicles, the up-tempo music slows with new car thefts, and Bonnie steadily abandons any hope of transforming their freedom into something sedentary and real. In effect, the declining fortunes of the gang, combined with Bonnie's sense of impending calamity, communicate the transience of the "freedom" they have attained. The bright possibilities of the road and subversive autonomy are exposed to be only temporary successes against the repressive forces of mass society. In the end, we see none of the fallen heroes, only their bullet-riddled car.

STOPPING THE PROFESSIONALS?

In response to the post–World War II crime wave, the FBI coordinated its efforts with local and state authorities, but the FBI's major endeavor was devoted to the organized-crime side of the auto-theft industry. With the FBI laboratory's cooperation, VIN and engine-block alteration could be detected using the agency's metallograph. And the FBI also shared information from its national automotive paint file that correlated make and model with color and pigment. One visitor to the laboratory in 1949 described FBI scientists at work:

> I stood in a room about 20 feet square. The blinds were drawn and the room was dark, except for a single light over the center of a workbench. Its powerful rays focused on two square magnetic terminals from which thick cables ran to a black switchboard.
>
> An F.B.I. agent in a smock threw a lever and a heavy humming filled the room. . . . The assistant handed him a sample bar of cast iron about a foot long which had been cut from a larger piece of metal.
>
> "Notice there isn't a mark to indicate where the stamped number has been ground off," the agent said. I ran my fingers over the metal. There was not a scratch.
>
> "Now watch," he said, placing the bar between the terminals. The humming rose to a whine. The assistant picked up a rubber hose and poured an oily, red fluid on the metal. "Magnetic oxide of iron in oil solution," he explained.
>
> Thin curls of blue smoke arose from the ends of the bar where they touched the terminals. The agent pointed. I saw nothing but

soupy red riffles on the bar. Then, suddenly, numbers appeared, faint at first, then clearly visible. There they were: 456-431. The agent scribbled a note and turned off the current. The numbers vanished.

"When the serial number was stamped on the metal, molecules were disturbed," the agent explained. "This magnetic flux sends a powerful electric current through the bar, and the impulses are interrupted en route by the same molecular disturbances. You might say the impulses are forced to detour, upward and downward. In doing so they agitate the magnetic oxide of iron we poured on the bar. The agitation takes the outline of numbers making them visible."[36]

The demonstration of scientific instrumentation impressed the journalist, but in fact it did little as a quick fix. Car-theft rings were big business, and it often took luck, happenstance, or human frailty rather than systematic science-based criminology to break the work of a well-organized gang. Particularly during early post–World War II years, cars were in short supply, both in the United States and in South America and Europe. Auto-theft rings proliferated in America, with memberships ranging from three individuals to twenty or more. One organization exported between three hundred and four hundred cars a year in 1949. Between the late 1940s and early 1950s the *New York Times* was replete with articles of smashed auto-theft rings, centered not only in New York City but also elsewhere.[37] For example, one ring was located in the South and specialized in 1949 and 1950 models, while a gang in California consisted of ten air force privates who stole only new cars. Often, key figures in these organizations were used-car dealers, perhaps because they had access to bills of sale and other materials that could be used as proof of ownership. Other important gang members who were garage attendants could acquire duplicate keys, and of course mechanics and car painters played a crucial role in vehicle alterations. The foot soldiers behind these operations were everyday people, including butchers, cab drivers, electrician's helpers, dock checkers, railway policemen, musicians, and housewives. And notwithstanding all of the science at the FBI's disposal, these gangs were often discovered after a chance traffic violation or when someone noticed a report of a car for sale well below the market price.

Organized car theft was also potentially violent, and this aspect was captured in Clyde Edgerton's regional novel set in 1950, *The Bible Salesman*.[38] This work is a humorous, homespun, Piedmont tale, in which car theft is one focus. The story is about a twenty-year-old Bible salesman named Henry Dampier, and the place is the hardscrabble red-clay South, with its marginal poor and those who prey upon them. Henry, a con man in his own right, obtains free Bibles and then sells them. But as it turns out, he is conned by someone much more experienced and at times violent—Preston Clearwater, who is a part of a car-theft gang. Attracted to Clearwater by the latter's new 1950 Chrysler, Henry becomes a driver of the Chrysler and then later of stolen cars. He knows the cars have been stolen, but Henry thinks that he is actually doing good, convinced by Clearwater that he is working for the FBI in an effort to break up car-theft rings! Car theft is definitely a part of this novel, but only peripherally, as the author's emphasis is far more on faith, love, doubt, and human foibles. Clearwater and his cronies employ more muscle than guile, and only chop-shop operations are described in any detail. If one learns anything about auto theft from this work, it is painting operations at the chop shop. The application of a special rubbing compound to a freshly repainted car produced a patina, thus decreasing possible suspicions that the worked-over car was hot. In the end, Henry is absolved of guilt and rewarded by finding love, while Clearwater meets a violent end.

Although juvenile joyriders were responsible for the majority of auto thefts during the 1950s, during every decade of the twentieth century, organized gangs played a significant and at times a leading role in this criminal activity. The transition from joyrider to hardened criminal is best illustrated by table 2.10, which indicates the diminishing percentages of juvenile arrests and the increasing percentages of adult arrests for auto thefts. The 1956 film *Hot Cars* opens with handsome car salesman Nick Dunn (John Bromfield) demonstrating a Mercedes 190SL to the beautiful Karen Winter (Joi Lansing).[39] Despite his charm and a drink, Winter drives away without purchasing the vehicle. Later that evening Dunn again fails to close a sale, this time with prospect

Arthur Martell (Ralph Clanton). Dunn lets his customer know that the MG-TD he has been looking at was previously in a rollover wreck, and by doing so ostensibly impresses Martell by his honesty. Dunn's failures lead to his firing at the end of the night, but the next day Martell summons him to his office, reveals to the salesman that he is in the car business and was just scouting the competition the night before. By offering the now-unemployed but honest Dunn a job, he takes the beleaguered salesman out of a tight financial spot, especially so because Dunn's young son is grievously ill and in need of a major and costly operation. The job with Martell turns out to be a bust, however, because in reality his car business is a front for a multistate stolen-car operation. Yet, after initially quitting, Dunn returns to the job because of his son's health issues, knowing full well that it is a dirty business. But it pays well, and Dunn is a family man in financial distress. With his life now sold away, Dunn quickly rises in the organization and gains enough trust that he is taken to the "refrigeration plant," where hot cars are moved to "cool down." There serial numbers on motors are filed and cars are given new paint jobs and reupholstered. Dunn is informed that papers are no problem either, for a printing and engraving business exists within the Martell organization. There is one fly in this ointment. State police investigator Davenport (Dabbs Greer) starts snooping around the car lot and later gets too close to breaking the ring when he attempts to buy a hot car.

As ring leader Martell states to Dunn, stealing cars means "not hurting anyone," since insurance covers the owner's loss and the "buyer gets a good deal." But when Detective Davenport is murdered, Dunn is implicated. It is Smiley Ward (Mark Dana) and not Dunn who did the killing, but that is not evident to the police. Dunn, who has a loving wife (and also forgiving, as it turns out), is without an alibi after a one-night affair with Winters, who all along was in the employ of Martell. As the investigation unfolds, the lid is finally blown off, and Dunn is seen for what he truly is, a pawn in a complex game. A subsequent confrontation between Dunn and Martell's assistant Smiley Ward results in a spectacular fight on an amusement-park roller coaster, with Ward plunging to his death and Dunn ultimately vindicated. Whatever its shortcomings as a B-grade film with a largely unknown cast,

Hot Cars portrayed car theft as a crime involving hardened criminals who were not averse to resorting to violence when forced to do so. Car theft has victims up and down the line and is more than victimless; it is an evil that touches on behavior beyond a direct connection with the automobile.

In reality, the head of a ring was rarely brought to justice. It took from the 1930s to the 1950s to end the career of "the king of car thieves" Gabriel Vigorito (a.k.a. Bla-Bla Blackman). Vigorito was serving a ten-year sentence during the 1930s for violating the Dyer Act when, with letters of support from New York congressmen and Senator Royal Copeland, he was given a conditional parole in 1939.[40] The intervention of Senator Copeland is most puzzling given Stephen Fox's characterization of Copeland in *Blood and Power: Organized Crime in Twentieth Century America*. Paradoxically, Copeland conducted the first congressional hearings on organized crime in 1933! Fox describes Copeland as "hard to place in the political landscape: a big-city Democrat, but conservative; a county doctor, always friendly and helpful in personal contacts, but tied to the Hearst and the Tammany machine."[41] After his 1933 anticrime campaign, however, Copeland returned to other interests, particularly to pure food and drug legislation. His support of Vigorito raises the question of whether Copeland the crusader ultimately became Copeland the corrupted by the end of the 1930s.

Despite the support of many citizens of New York City and Brooklyn who wrote on his behalf, only a few years later Vigorito was in trouble again, charged with illegal trafficking in untaxed alcohol, extortion, coercion, and the selling of black-market gasoline. Beginning in 1948, Vigorito and his gang stole Cadillacs and exported them to South America and Europe. It was only when FBI agents infiltrated the gang that the full extent of its operations emerged. Later, the *New York Times* reported that Vigorito's operation had handled more than $1 million in stolen cars in 1952.[42] Since Vigorito was personally isolated from the operation, it was difficult to pin illegal behavior directly on him. Finally, his fortunes faded after 1954, when forty-nine witnesses lined up to testify against him.[43] However, even then, along with three of his sons and two nephews, "Bla-Bla" still managed to run his ring from Fort Leavenworth prison.[44] Finally caught writing letters directing lieutenant "Buck" Murray, Vigorito spent the remainder of the 1950s tightly locked up, ever proclaiming that he was "framed."

As Vigorito served his time, other masterminds or kingpins working independently quickly replaced him in the New York City metropolitan area with operations of their own. Harold Wapnick, a small-time money lender when he was first arrested for auto theft in 1956, was one such businessman. Unlike Vigorito, whose only education had come from the street, Wapnick was an accountant with degrees from St. John's University and Columbia University.[45] By 1962 Wapnick had put together an acquisition, modification, and distribution production line that was described in the *New York Times* when Wapnick was sent to jail:

> At one time, Wapnick had sixteen men working for him. In the six-month period before their arrest, the ring's members disposed of at least sixty cars. The Federal Bureau of Investigation and city detectives said the gang handled 20 per cent of all the cars stolen here.
>
> The ring operated in this way. Charles Gersh, Jose Monteiro and MacAlan Gladding ran junk yards where late-model wrecked cars were stripped of their identifying numbers and license plates. George Dooley, Dominic Schiavo and Walter Bedell stole cars corresponding to the models in the yards.
>
> Philippe Roy and Harmon Gripper ran an outfit called Automotive Reconstruction Enterprises, where the stolen cars were equipped with the identification of the wrecked vehicles and repainted to conform with them. Then James LaFazia and David Brill delivered the cars to points in Connecticut, Pennsylvania, Massachusetts, New Jersey, and California.[46]

News stories about organized auto-theft rings abounded during the 1960s and 1970s, and often there were common elements associated with each case, regardless of what place the thieves called home.[47] While this feature was not new, some rings stole cars to order, and in one case their customers included a professional football player, an entertainer, and a screen star. Cadillacs and Lincolns were the preferred targets, although one New Jersey gang specialized in Volkswagens.

Another gang stole lemons, by arrangement with their owners; this served as a remedy before lemon laws were enacted. Particularly during the 1960s, there were accusations of mafia involvement in the auto-theft business, usually involving lower-level functionaries tied to major crime figures including the Columbo family. Partly as a result of the construction of interstate highways during that decade, the radius of operations expanded considerably, and in the case of the New York City area, Florida became a major destination for stolen cars. Police were integral members of some of these groups, as were Department of Motor Vehicles employees who processed new registrations. Finally, stripping cars was commonplace, as thieves worked with wrecking yards to crush the body shell, thus destroying the most important piece of evidence.[48]

Interstate highways proved to be analogous to the great rivers in nineteenth-century America. To reduce the flow of hot cars between states, police departments and the insurance industry introduced electronic communications and computers during the 1960s. The teletype machine, first used by law enforcement in 1927, was tied to local and then regional networks in 1963. Initially, teletypes offered limited deterrence because of the lack of a standardized data format, but in 1965 a nationwide information system was proposed that in 1966 became the Law Enforcement Teletype System. Western Union machines used punched paper tape, transmitting license and registration information twenty-four hours a day, seven days a week. However, as more and more users availed themselves of the system, gaining timely access to it became a problem. Switching devices located in Phoenix, Arizona, were the first step in developing a more elaborate system, which was proposed in 1966 by the FBI and the International Association of Chiefs of Police. The following year, two IBM 360 computers housed at the FBI's National Crime Information Center were put into use, containing records that could be accessed on demand. The NATB's 3×5 card system located in Chicago was then translated onto eighty-column punch cards. The NATB had developed its own computer capabilities by 1970, starting with an IBM 20 machine and then upgrading to an IBM 370 in 1973. This development resulted in the North American Theft Information System, a technology that allowed NATB regional divisions to communicate via cathode ray tubes. Information could now be sent to offices in seconds, and such speed was critical, given the rise in auto thefts during the 1970s.

RAISING AN ALARM

The wave of auto thefts in the early 1970s and the failure of manufacturers to make vehicles that were secure prompted the rising popularity of aftermarket security alarms. A wide variety of security alarm devices were available for virtually every make and model of car, ranging in price from about $30 for owner-installed devices to about $130 per unit for seller- or factory-installed alarms. In general, a security alarm system consisted of a control unit, a set of sensors that might include pin switches installed at a door, in the trunk, and under the hood; a motion sensor; a switch to cut off the starter motor; and a siren or horn to indicate an attempted theft. More elaborate designs involved current draw, wheel rotation, or ultrasonic detectors.

Electronic alarms were not new; the concept had begun to occupy inventors as soon as they thought such a solution to auto thefts was technologically feasible. For example, a June 20, 1920, article in *Popular Mechanics* described a loud alarm devised by a Nebraska inventor: "[It] utilizes the drive-shaft to operate its own bell or horn signal when the car is improperly moved. A friction gear, thrown into or out of engagement with the shaft by a cam, is enclosed with the alarm in a riveted steel case, fixed to the shaft housing and radius rods. The cam also short-circuits the magneto, so that turning the key in the lock stops the engine and sets the alarm. The lock is located in the floor of the driving compartment."[49]

The mass-market auto theft alarm, however, appeared only after considerable development and miniaturization, made possible after the introduction of solid-state electronics. The post–World War II design of automobile alarms began with Victor Helman's "Automatic Burglar Alarm," patented in 1954. Helman, from Cleveland, Ohio, developed a technological system that included a secure, resettable control box that was connected to switches securing the doors, the hood, and the trunk of a car. If unauthorized use was detected, two electromagnetic solenoids would trigger a current that energized a siren. Further technological improvements to the concept of a car alarm were developed later in the 1950s. For example, Joseph Yurtz, also a Clevelander, came up with an automobile theft alarm that was based on a motion sensor. Two vibrators, placed in such a way that they were perpendicular to each other and thus could sense motion in two planes, would activate a horn "upon the slightest agitation of the vehicle." In 1971 inventor Charles E. Davis solved the problem of battery drain, which occurred with alarm systems that either would always be on or that would continue to sound off after activation.[50]

One auto alarm kit, the 1967 Auto Sentinel, was claimed to be so effective that "the only way your car can be stolen . . . is for the thief to pick it up bodily and carry it away."[51] For a modest fifteen dollars, one could construct a device consisting of several relays, a switch, and a pair of resistors that would be triggered by mercury switches or existing door switches that were connected to the horn relay. For the would-be owner, all that was left to do once the Auto Sentinel was mounted and alarmed was to "relax"; unfortunately, all the thief had to do was to find the box and disconnect terminal 1!

Several years later, the Radatron Corporation produced an auto alarm kit that was designed for beginners to construct, containing a fifty-two-page instruction booklet that concluded with the question "What did you learn?" Employing a power transistor and a control box with five coded push buttons, the unit caused the vehicle's horn to sound intermittently when it was triggered by any electrical signal. The alarm was a deterrent at best; and the manufacturer warned any would-be kit builder that it "does give protection but should not be expected to foil a professional car thief familiar with this type of alarm."[52] Do-it-yourself electronic alarm kits of that era also often included motion sensors based on a pendulum. Unlike a clock pendulum, this one was designed to be still rather than to swing all the time. Hanging straight down, it made no contact and thus the alarm was silent. If someone opened a door or leaned on the car, however, the pendulum swung on the low side, making an electric contact that resulted in sound.[53] The weak point in all of these primitive alarm systems was simply the power source used to energize them. All a thief had to do was to cut the battery cable to the car; once that was done, the alarm was dead.

Chrysler Corporation offered a more sophisticated system in 1973 that certainly was aimed at deterring at least the marginally skilled professional thief.[54] Available on all full-sized Chrysler vehicles, the Chrysler Electronic Security Alarm System featured an integrated-circuit electronic control box coupled to a series of sensors located in various areas of the car. When the sensor was tripped by a forced door, trunk, or hood, or an unauthorized start, a pulsing horn and flashing lights indicated that something was wrong. The Chrysler system also featured a panic button on the dash that automatically locked all doors, as well as energizing the horn and lights. The key to the door activated the system and deactivated it as well. It was a system that reflected the concerns of a new age in American life, one in which unlocked doors and trusting glances were replaced by suspicion and fear. Further improvements evolved during the late 1970s as electronics got better and better. Silicon-controlled rectifiers and integrated circuits could be combined in such ways as to design sequential digital code systems; unless all four numbers on a box were pressed properly, the alarm system was still alive. Yet, this too could be circumvented, as auto thieves often proved more ingenious than circuits.[55]

Factory-installed systems from the early 1970s were far from effective. In the case of the Corvette, for example, an alarm was installed beginning in 1972. It did little to stop the seasoned professional thief, however, as reflected in the Senate testimony of "X," a specialist in stealing Corvettes between 1966 and 1977: "Well, all Corvettes after 1972 have an alarm. I would walk up and look at the fender to see if the alarm was turned on, I would reach under the back, by the bumper, and pull the wire out of it. That is how easy it is to disconnect the alarm."[56]

One popular device available during the 1970s and 1980s was sold by Chapman Security Systems of Elk Grove Village, Illinois. David Arlasky, the founder and president of Chapman Industries and the inventor of the CHAPMAN-LOK, was a former auto thief. According to a company brochure, Arlasky purportedly knew "all the tricks of the trade in car thievery. For almost two years, he was a car repossessor for Chicago Banks. He 'stole' nearly 2,000 automobiles and made enough money to put himself through college! 'And professional car thieves make a good deal more than I did,' he says about the motivation behind auto theft."[57] Technologically, the Chapman system consisted of a key-operated hood and ignition lock placed under the dash. Protected by a steel conduit that was strong enough to resist a bolt cutter, a wire was fed under the dashboard to the automobile's distributor, while a cable activated a dead-bolt lock firmly secured to the hood. Optional accessories included a flasher, a panic button, a horn honker, a sonic sensor, and a glass sensor.

Alarm systems deterred the joyrider but not the professional auto thief, who had the expertise to cut the right wire, thus disabling the system. And by the late 1990s, the alarms' effectiveness was challenged by both the insurance industry and civic activists, who increasingly complained of false alarms. Statistics for 1997 tended to conclude that there was no difference in theft losses based on having an alarm system installed in a vehicle. Furthermore, critics cited the statistic that some 95 to 99 percent of alarms were false, supporting the notion that car alarms merely degraded the urban environment. To reinforce the contention that alarms did little to reduce the crime of auto theft, the Progressive Insurance Company claimed that only 1 percent of citizens would call police if they heard an alarm.[58]

Thus, over the span of two decades and despite their increasing sophistication, alarms fell out of favor as a theft deterrent. Thieves were just too clever, no matter what kind of electronics were used. Thomas Bowman, the insurance director of the Automobile Club of Michigan, confirmed Lee and Rikoski's work in a 1984 study where he concluded, "There can be no doubt that the entry of organized crime has driven our rate of recovery of stolen vehicles down to 54 percent in 1982 from a rate of almost 83 percent in 1974."[59] However, these low recovery rates proved transitory; they improved dramatically by the late 1990s, in part owing to technologies that included the LoJack, and in part the result of a strong federal government response, the cooperation of manufacturers, regional policing, innovative insurance industry methods, and defensible spaces.[60]

3

From the Personal Garage to the Surveillance Society

Terrified by crime and worried about property values, Americans are flocking to gated enclaves in what experts call a fundamental reorganization of community life.

PLANNING 60, 1994

How we create our physical environment—our houses, cityscapes, open areas, streets, and sidewalks—says a lot about us as a society. Whether we are orderly or chaotic, inviting or insular, an advanced economy or a developing one are all expressed in the way we organize our space. And when our values change, they are similarly etched in our landscape as a reminder of what we were and what we are becoming. Such is the case with auto theft, particularly in the era after World War II. For a long time, we didn't design our built environment with deterrence in mind, but now we do, a reflection of an increasingly violent and fragmented urban society.

In retrospect, Americans could easily have incorporated the deterrence of car theft into the built environment at the time the automobile was introduced and throughout its early years, but that did not happen. As cars gained widespread usage, theft was not a primary concern in creating either the transportation infrastructure or personal spaces. On the contrary, designing out crime through walls, gates, and surveillance was until recently associated with countries plagued by widespread lawlessness, extreme economic disparity, and political upheavals—not the United States. For much of American history, we have thought of our society as socially stable. However, times have changed. Since the 1970s, discussions about "defensible space" and "crime prevention through environmental design," which includes stemming car theft, have become common. This dialogue has also become emotionally charged, since it is often linked to poverty and race. Americans are discovering that the physical and social barriers aimed at fighting crime force a trade-off between personal security and individual freedom, making this a conflicted choice.

So what changed to make environmental design a viable approach for enforcing laws and reducing crime, including auto theft? First was a rise in offenses during the 1960s, 1970s, and 1980s, when the overall crime rate in the nation soared despite an often-strong economy. Auto theft alone grew by 56 percent from 1970 to 1990. But then, over the next two decades (1990–2010), crime as a whole dropped by a significant 28.6 percent, while auto theft showed a dramatic decrease of 51.4 percent. Even though crime (including auto theft) has been on a steady decline, fundamental changes in our society and economy have fueled a lingering and perceived breakdown in order.

Beginning in the 1960s, people in the United States began to feel that their lives were more fragile. Changing urban demographics led to what has been characterized as "white flight." Immigrants and minorities moved into metropolitan areas as the majority population took to the suburbs. The end of the Cold War and the rise of the global economy led to a massive shift in the nation's industrial landscape throughout the 1980s and 1990s. As jobs were off-shored from the United States to other countries, middle-class incomes eroded. Then in 2001, Americans became victims of the worst terrorist violence on U.S. soil, followed by a financial crisis in the fall of 2008, the collapse of the housing bubble in 2009, and the highest nationwide unemployment rate since the Great Depression.

The high crime rate characteristic of earlier periods, a new era of terrorist assaults, and the climate of heightened uncertainty generated an underlying desire for safe environments. New housing developments proudly touted strengthened security, while public spaces were modified to offer increased protection. Americans began the uneasy adjustment to a new normal. The effort to stop thieves from stealing cars was never the main objective of the changes that created "fortress America," but it was a bellwether.[1]

How the prevention of car theft is reflected in the built environment can be seen in design changes in the home, including personal garages and gated communities, and in public spaces such as streetscapes, parking lots, and structures.

THE HOME AND THE AMERICAN GARAGE

Common sense might argue that early garages were built to prevent auto theft. In reality, they were primarily places for inventing, constructing, and rebuilding cars and protecting them from the elements. Pioneer Brass-Era cars were notoriously loud, dirty, dangerous, and most significantly, very expensive. They required constant maintenance and needed to be housed in buildings large enough to hold parts and equipment. In addition, the buildings needed to be sufficiently distant from residences, in case of an explosion or a fire (the most frequent reason for early auto-insurance claims). These structures were converted carriage houses, barns, and sheds—outbuildings dedicated to transportation that were sometimes shared with horses and buggies.[2] One famous early garage, sometimes considered the "first" garage, was Henry Ford's workshop, situated behind his duplex at 58 Bagley Avenue in Detroit. He constructed the 1896 Quadricycle in that workshop. When the contraption was completed, it was too large to fit through the existing door, so Ford hammered and chiseled a way out, thereby creating what was arguably the first garage door.

Cars started out as a novelty in the late 1890s but had become a necessity by the 1920s and 1930s, because of the growth of suburbs. Concurrently, interest in household garages grew; a place was needed to protect these sometimes temperamental but prized objects. A 1913 article in *Collier's* magazine cautioned that "a car stored in an unheated space could end up with a frozen radiator, rattling doors, bowed fenders, a cracked frame, and flaking paint."[3] A heated garage was a luxury few could afford, but an enclosed space was not. By the 1920s, even though most people did not own a garage, it was a dream to which they aspired.[4] President Herbert Hoover knew this when he promised "a chicken in every pot and a car in every garage" during his 1928 presidential campaign. The slogan struck a chord that resonated in the

American psyche. Throughout the Great Depression, detached and attached garages began popping up across the country.

For the most part, the focus during the 1930s was on whether to build an attached garage, a semi-attached garage, or a detached garage, with an emphasis on architectural integrity or convenience, but theft was rarely a consideration.[5] In fact, although garages were desirable, home-buyers did not always select a home with a garage. From 1933 to 1940, Sears, Roebuck and Company offered home plans in their catalog that gave people choices—they could build a garage, or a dining nook, or a recreation room or the like, according to their preference. The company also offered freestanding garages and pergolas for sheltering cars that could be bought separately and built later.

With garages came garage-door-opening systems for both convenience and, later, security. The garage-opening devices of the 1920s sometimes used a spring beneath the driveway that was activated when the driver entered, or a chain or rope next to the garage door that the driver pulled to open the door.[6] However, the spring doors could injure a family member who happened to be in the garage, and neither mechanism adequately secured the automobile.

By the 1930s, companies began to consciously aim at preventing theft, using photoelectric cells or radio signals for the purpose. These ingenious early systems were still uncommon and were forerunners of the widely adopted methods of later years. *House & Garden* described the technology available at the time: "A photo-electric cell installation . . . opens doors when [a] light beam . . . is broken. . . . [Alternatively, a] remote control on [a] standard in [the] drive[way] . . . operate[s] doors electronically when unlocked, by key, from car. Both from Stanley Works. There is also a radio type door operator, from Barber-Colman, which works by short wave from car when you pull knob on dash."[7]

Remote-control garage-door openers of the 1930s generally consisted of a simple transmitter and receiver that controlled the mechanism. One type, developed in Spokane, Washington, used a receiver that responded only to a particular wavelength broadcast by the spark coil on the car. Another system, developed in Illinois, used coded radio signals sent from an instrument board in the car to an antenna buried in the driveway, which started the electric motor that opened the garage doors. Although automobile thieves could potentially find the right radio wave to thwart the Spokane system, they were hard-pressed to crack the Illinois designers' radio code.[8]

By the 1940s, consumers also turned to garage-door-opening schemes like that of the Chicago-based Era Meter Company's "Drive-Rite-In." It opened the garage door via a key-operated post at the end of the driveway.[9] But for the most part, these and other devices were intended to make it easier for the driver to enter the garage.[10] Whether detached, semi-attached, or attached, garages often had separate entry doors with glass windows, and it was very easy to break in.

After World War II, the attached garage became a common feature of the typical American suburban home. The 1960 U.S. Census enumerated garages for the first time. With advances in technology, garage-door openers were also becoming more sophisticated. The key system gave way to electronic garage-door openers. These devices of the 1950s and 1960s used a simple code method that operated the garage door with a push of the remote's button. However, the remote controls had a "shared frequency"—a garage door was opened by one of five codes—and that was problematic in its own right. Because there were only five codes, a suburban driver could easily have the same code as his neighbor. Thus, the goods inside the postwar suburban garage, and indeed even in the home itself, were easy prey for a criminal.

By this time, garages were being used as transitional spaces that linked the indoors to the outside. They were refuges for playing music, working on hobbies, storing belongings, washing and drying clothes, and puttering, and driveways were convenient playgrounds for shooting hoops and roller-skating.[11] With few exceptions, people in Middle America did not lock their garages; on the contrary, garage doors often stayed open day and night.

However, as crime rose during the 1960s and 1970s, many Americans began locking their garages, and companies that sold garage-door openers began to devise more secure systems. By the 1980s, companies equipped their remote controls with dual in-line switches that sent between 256 and 4,096 codes from the remote control to the garage-door opener. Criminals were still able to hack this system, however, by trying multiple codes on a regular transmitter or using code grabbers to attempt every possible combination in a short period of time. It was not until the late 1990s that garage-door-opening systems used "rolling codes" that changed with each opening of the door, making it almost impossible for thieves to reproduce the code.[12]

With the progressive changes in garage-door technology and the loss of the innocence associated with leaving the garage door open all night, the garage and its contents became relatively safe from theft. Automobile-insurance underwriting recognized the effectiveness of the changes in stemming auto theft. Although the individual home and garage could provide significant protection against auto theft, for some this was not enough. For those who needed additional assurance, gated communities that were walled off from the rest of society offered even greater protection from crime, including auto theft, and their appeal has grown from the 1990s to the present.

GATED COMMUNITIES

Another strategy to secure the car was to place it in a gated community. These safeguarded neighborhoods, which offer enhanced security for a resident's car, have a long history both in the United States and abroad. They were initially created as exclusive estates and later designed as retirement or resort communities. They now tend to focus on middle-income people who seek a sense of safety and security. National interest in gated communities intensified during the late 1990s, even though by that time crime had peaked and was actually decreasing. The number of people living in gated communities in the United States was estimated to have doubled from 4 million in 1995 to 8 million in 1997.[13] The International Foundation for Protection Officers suggested that this represented approximately twenty thousand gated communities, or 3 million units.[14] By the time of the U.S. Census Bureau's 2001 American Housing Survey, more than 7 million households, or 6 percent of the national total, were located in developments delineated by walls and fences and with controlled access. About 4 million of these communities are further distinguished by self-governance.[15] In these cases, gated communities are defined as "multi-unit, master-planned

developments where resident owners must be members of a home-
owner association and share ownership of common facilities, including a surrounding fence with a gate."[16]

The popularity of self-governed gated communities has also been
interpreted as the ascendance of a movement toward privatization that
began in the 1980s in response to a half-century of public-sector expansion. Gated communities are a personal example of the privatization
of security, a realm previously relegated to the public sector along with
the maintenance of peace. In his analysis of San Diego's gated communities, Sven Bislev of the Copenhagen Business School suggests: "The
idea of self-governance could be said to be 'postmodern,' in the sense
that instead of overarching models, grand narratives, and historical
schemes, it privileges individual and situational choice. This, however,
also excludes the possibility of something radically different."[17] That is,
gated communities provide a form of self-governance at the neighborhood level aimed at averting risk, promoting common sociocultural
values, and providing an alternative lifestyle to people with means at
a time of growing economic disparity. These themes continue to be
played out in different arenas and are at the heart of political debates
that frame contemporary national discourse.

Although gated communities are now common across the United
States, the highest concentrations are in the Sunbelt. About 30 to 40 percent of new homes in California are in gated communities, as are most
recent subdivisions in Palm Beach County, Miami, and Tampa, Florida;
Phoenix, Arizona; and Washington, D.C.[18] It is reported that eight out
of every ten new urban developments across the country are gated.[19]

Gated communities use a variety of methods, from gates to cameras, to achieve both expressed and implied exclusivity. It is a lifestyle
choice. Edward J. Blakely and Mary Gail Snyder describe the security
features:

> Gates range from elaborate two-story guardhouses manned 24
> hours a day to roll-back wrought iron gates to simple electronic
> arms. Guardhouses are usually built with one lane for guests
> and visitors and a second for residents, who may open the gates
> with an electronic card, a punched-in code, or a remote control.

Some gated communities with round-the-clock security require all cars to pass the guard, and management issues identification stickers for residents' cars. Others use video cameras to record the license plates and sometimes the faces of all who pass through. Unmanned entrances have intercom systems, some with video monitors, for visitors seeking entrance.

These security mechanisms are intended to do more than just deter crime. Both developers and residents view security as not just freedom from crime, but also as freedom from such annoyances as solicitors and canvassers, mischievous teenagers, and strangers of any kind, malicious or not. The gates provide a sheltered common space that excludes outsiders. Especially to the residents of upper-end gated communities, who already afford to live in very-low-crime environments, the privacy and convenience of controlled access are more important than protection from crime.[20]

Gated communities are never marketed as lowering car theft per se, but rather as offering relief from a crime-ridden society. Even though their effectiveness in lowering crime has been spotty and difficult to document, car theft is one crime that has actually shown some temporary improvement. According to one 1995 study, "in Miami and other areas where gates and barricades have become the norm, some forms of crime, such as car theft, are reduced. On the other hand, some data indicate that the crime rate inside the gates is only marginally altered."[21]

Not surprisingly, car theft is sometimes mentioned as a reason to move to a gated community, since this crime hits close to home. Anthropologist Setha M. Low, in her studies of gated communities, captured the sentiments of people who made the move. One New York resident said: "I think it's safer having a gated community. . . . They are not going to steal my car in the garage. . . . [In the old neighborhood] every time we heard an alarm, we were looking out the window. My daughter and son-in-law lived next door and their car was stolen twice." Another stated: "I got to feel like I was a prisoner in the house. . . . You didn't park on the street too long because you are afraid your car is going to be missing something when you get out, or the whole car is missing. . . . So there's a lot of things we have the freedom to do here that we didn't do before."[22]

What is it about car theft that people fear? Like personal assaults and home invasions, car theft feels like an attack that plays on a person's vulnerability. Car theft triggers an emotional reaction that is impervious to statistics that claim crime across the United States is actually on the decline. Typically, a person whose car has been stolen feels so violated that the car, after it is recovered, is later sold.[23] Sociologist Barry Glassner suggests further that we overreact because of a barrage of media reports glorifying crime and violence, which create an image of a crime-ridden society, although that image is, in fact, disconnected from reality.[24] Sensationalized headlines both feed off and incite a "culture of fear."

Anthropologist Setha Low, urban planners Edward J. Blakely and Mary Gail Snyder, and social critic and political journalist Barbara Ehrenreich are among the many voices contending that gated communities play on this fear when they implicitly and explicitly delineate people by race, income, and social standing. Residents don't have to mix with immigrants, minorities, the poor, and young delinquents whom they associate with urban change and a chaotic world. In the words of one resident: "When Bloomingdale's moved out and Kmart moved in, it just brought in a different group of people . . . and it wasn't the safe place it was. . . . I think it's safer having a gated community."[25]

Gated communities are also a way for developers to appear to offer a private good to supply what the public seemingly has failed to provide, while making money from the strategy.

> After giving a short presentation Monday [December 9, 2009] to the Charlotte-Mecklenburg Planning Commission, planning commissioner Nina Lipton asked the Charlotte-Mecklenburg police chief whether he had any data on safety in gated versus non-gated communities.
>
> "We looked at that," [police chief] Monroe said. The police and planning departments matched up communities as closely as they could, looking at income levels, multi-family, single family and other factors. In terms of crime rates, Monroe said, "We saw no difference."

What matters in terms of neighborhood safety, he said, is who's living there: Are residents looking out for their neighbors? Are they taking responsibility? Is it a rental community, is there professional management? Are renters being screened for criminal records?

. . . Just making a development gated doesn't make it safer, he said. "Sometimes it creates an opportunity for me to charge you more."[26]

Gates and video surveillance systems offer a sense of security to some, but to others they foster a false sense of community. As a tool for preventing auto theft, gated communities can even have the opposite effect. Neighbors may become complacent about their environment and exercise less vigilance than they would without the barriers. Equipment or cameras may malfunction, gates may be left open, or security guards may be distracted. Because crime prevention cannot be guaranteed, there is reportedly a recent trend in the multi-housing industry to avoid use of the word *security* in their marketing material.[27]

Could we actually be seeing the beginning of disillusionment with gated communities? Ehrenreich claims that gated communities are "another Utopia [that seems] . . . to be biting the dust" and reports that "America's gated communities have been blighted by foreclosures," as evidenced by what has occurred in Henderson, Nevada, and Orlando, Florida.[28] Gated communities may be on the wane, but they are not going away, regardless of whether their allure for preventing crime is losing its luster. Their appeal based on protection from outside influences, even if largely symbolic, remains strong. And as long as local law enforcement budgets are constrained by a soft economy, and as our society ages and the allure of "self-governance" prevails, a culture of fear will continue to play on the American psyche.

MAKING PUBLIC SPACES SAFER

The 1950s and 1960s in the United States marked a period of concern over growing poverty and social inequality. The War on Poverty sparked the building of large urban-renewal projects that, in turn, initiated a lively debate that continues to this day over the relationship between human behavior and the built environment. The question among those interested in crime prevention was and still is whether good urban design can reduce crime, including car theft. Activist Jane Jacobs argued in her book *The Death and Life of Great American Cities* that successful city neighborhoods, through their construction of spaces, encourage ordinary citizens to be the first line of defense against crime. C. Ray Jeffery turned the concept into a multidisciplinary approach to crime deterrence in his *Crime Prevention through Environmental Design* (1971), and the approach was later popularized by Oscar Newman in his books *Defensible Space: Crime Prevention through Urban Design* (1972) and *Design Guidelines for Creating Defensible Space* (1976).[29] Critics complaining of "environmental determinism" forced a refinement of the ideas so that current iterations are more cognizant of site-specific and social factors.[30] In the wake of these heated discussions, organizations such as the United States Designing Out Crime Association, groups committed to implementing these design concepts, have emerged, as well as efforts to document the assertions as fact or fallacy. From these lively conversations, it is possible to gain some insight into the relationship between environmental design and auto theft.

What we have learned so far is that it is sometimes difficult to isolate the specific elements that directly affect auto theft because what is important is how these elements are arranged together to create the total event. And it is not easy to draw conclusions based on statistical analysis, given the rigid requirements of the processes.[31] Nonetheless, some features do seem to make a difference: lighting, surveillance, landscaping that does not obscure the view of crimes as they occur, and the presence of security guards or police.[32] In addition, people need to be aware of where they park. As studies show, parking a car in a domestic garage at night is safer than parking in a driveway, which, in turn, is safer than parking on the street.[33] But not everyone has a choice.

STREETSCAPES, PARKING LOTS, AND STRUCTURES

Parking lots and garages initially were built in the American city as a way to ease congestion; later they were employed as a developmental strategy to offset commercial growth in outlying areas. As geographer John A. Jakle and historian Keith A. Sculle note in *Lots of Parking: Land Use in a Car Culture,* the parking garage marked a transition from parking as a privilege to parking as a necessity.[34] And while safety had become an issue by the 1970s, auto theft was not widely discussed. One of the early design elements to gain attention with respect to crime was lighting, yet even here the focus was on personal assaults and property stolen from cars.[35] In a 1979 evaluation of eight street-lighting systems in the United States, no measurable effect on vehicle theft was found in the one location (New Orleans) where that aspect was examined.[36]

The problem is that, although it may seem obvious that lighting would reduce crime, and auto theft in particular, the evidence is mixed. On the assertion that more lighting leads to less crime or safer streets, there is disagreement among experts. One large, long-term study found that "better street lighting had had little or no effect on crime. . . . On the other hand, . . . the improved street lighting was warmly welcomed by the public, and . . . it provided a measure of reassurance to some people."[37] In other words, improved lighting reduces fear. However, another study found something different: "Improved street lighting led to significant reductions in crime [since] lighting increases community pride and confidence and strengthens informal social control."[38] That is, it encourages informal surveillance or vigilance among residents. In fact, there are both positive and negative consequences of improved street lighting. While better lighting might bring people out in the street at night, it can and does help criminals see what they are doing and scope out their exit strategy. In addition, harsh lighting creates shadows that can hide criminal activity. So numerous variables affect the relationship between lighting and crime:

> The effects of improved street lighting are likely to vary in different conditions. In particular, they are likely to be greater if the existing lighting is poor and if the improvement in lighting is considerable. They may vary according to characteristics of the area or residents, the design of the area, the design of the lighting, and the places that are illuminated. For example, improved lighting may increase community confidence only in relatively stable homogeneous communities, not in areas with a heterogeneous population mix

and high residential mobility. The effects of improved lighting may also interact with other environmental improvements, such as closed circuit television (CCTV) cameras or security patrols.[39]

It's not that improved lighting does not deter car theft; it is just that a direct link is not automatic. The ability to mitigate auto theft through lighting often requires combining it with other measures such as landscaping and, increasingly, surveillance.

Electronic surveillance had been used by the federal government and the military for years, but it became more widespread in the 1960s during the war against organized crime, and it spread to municipalities when Congress passed Title III of the Omnibus Crime Control and Safe Street Act of 1968. The purpose of the legislation was to define the proper use of electronic surveillance. The Electronic Communications Privacy Act of 1986 further allowed law enforcement to collect information for public safety. Although the act has been challenged, court rulings have argued that individuals do not have a reasonable right to privacy when in a public space, nor does video surveillance of public streets violate the U.S. Constitution's Fourth Amendment protection against unreasonable search and seizure.[40] Thus, the legal groundwork was already in place at the time of the World Trade Center bombing and the Oklahoma City bombing, both of which "raised public concerns about security . . . [and] made the video surveillance industry more acceptable to the general public."[41]

Two early instances of the use of public video surveillance were in Hoboken, New Jersey (1966), and Mount Vernon, New York (1971), but both systems were dismantled after a few years when they produced few or no arrests. The problem was that "many of these early systems were technically and financially deficient, and lacked local public support. According to a police officer, 'Cops weren't thrilled with the cameras.' Police staff often had to sit in a room to monitor the closed circuit television (CCTV) cameras, which frequently broke down."[42]

But the introduction of camcorder technology in the mid-1980s and digital technology in the 1990s expanded the range and coverage of surveillance. When linked to computers, cameras equipped with sensors that filtered out unrelated information provided extremely high-resolution images that could be stored in databases. In addition, with passage of the 1994 Violent Crime Control and Law Enforcement Act, the U.S. Department of Justice began funding such programs, as well as the use of Geographic Information Systems to map and track gangs and other criminal activities. It also established regional law enforcement technology centers that could, among other things, provide technical assistance in the use of surveillance technology.

Consequently, by the mid-1990s, municipalities and private companies began to erect electronic camera surveillance systems to monitor high-traffic areas and parking lots where cars were left for an extended period of time, such as airport parking lots, shopping centers, schools, and commuter lots. And they learned from international experience with such technologies. In one example, the use of CCTV in the parking lots of a British university led to a 50 percent reduction in automobile theft.[43] And the installation of CCTV security in the town of King's Lynn, Great Britain, in 1988 led to a drop in the theft of cars from fifty-two to zero by 1994, while theft from cars fell from fifty-six to one.[44] When the municipality of South Orange, New Jersey, installed seven CCTV surveillance cameras in parking lots, intersections, and parks in 1994, auto theft reportedly dropped by 40 percent.[45]

Nonetheless, not all municipalities were prepared for the financial and labor commitment that came with the new systems. San Diego used CCTV surveillance in 1993 across the heavily used Balboa Park and realized a significant decrease in all crime, including car theft. However, police were forced to stop the surveillance after nine months because the program required public funding.[46] Across the country, similar mixed results were reported. In 1993, in response to the spike in auto thefts in parking lots used by train riders into New York City, authorities installed security cameras and signage at a 210-space commuter parking lot in Freeport, New York. Yet, when the nearby municipality of Nassau attempted the same program, they found the construction and maintenance of video cameras both expensive and labor intensive.[47]

Declining budgets forced communities to find ways to overcome the high cost and labor issues related to surveillance systems. The San Diego Police Department, for example, used cameras selectively and advised drivers to "park in open, well-lighted, and popular areas near [their] destination, preferably . . . in view of a security camera."[48] When there were successful results from a Hollywood, California, initiative by building owners and landlords to purchase and install CCTV cameras along a crime-ridden corridor that was monitored twenty-four hours a day by local volunteers and Los Angeles Guardian Angels, business tenants in the nearby Northridge Shopping Center pooled their resources to install sixty-four CCTV cameras in 1995 and realized "an immediate and sharp reduction in auto theft and burglaries."[49] After experiencing an increase in car thefts, the state of Florida gave the city of St. Petersburg a forty-two-thousand-dollar grant in 1994 to equip mobile police officers with mounted cameras and night scopes; the surveillance system reduced the thefts of cars within two months (the high had been three thousand) in the high-traffic "Gateway" between St. Petersburg and Tampa Bay. "According to police officials, the video surveillance combined with police officers with night scopes had a major impact."[50] Because the police were mobile and the cameras were not fixed in place, the security was extremely flexible.

Criminal justice scholars Ronald V. Clarke and Patricia M. Harris cautioned that, while promising, research on the role cameras played in preventing crime was inconclusive at best.[51] The effectiveness of the surveillance system was dependent on having people able to review the material in a timely manner. Consequently, when the California Research Bureau conducted a telephone survey of major city police departments to determine whether they would use CCTV video surveillance in public areas, most responded that it was not as effective as "community policing and other prevention strategies."[52] To what degree traditional policing methods were resistant to the new technologies, however, is a question that begs to be answered. Cameras have the potential not only to replace cops, but to monitor their activities, as well.

A detailed study on the theft of autos and their contents was recently undertaken by the Chula Vista, California, police department for the U.S. Department of Justice, Office of Community Oriented Policing Services. It provided not only comprehensive insight into how and why cars are stolen in this border city, but also how designing for crime prevention can come together in a focused effort.[53] Because Chula Vista is located only ten minutes from the Mexican border, it would be expected that the city would be a haven for auto theft. And, in fact, it was. Chula Vista had a much higher auto-theft rate than Los Angeles, New York, Chicago, San Diego, and Fort Worth, for example. A close examination of U.S. border cities shows high rates of auto theft in all of them, but Chula Vista's auto-theft statistics at 984 motor vehicles thefts per 100,000 in 2001 ranked higher than the figures for the Texas cities McAllen (670), Eagle Pass (424), Brownsville (374), and El Paso (326), yet lower than the rates for Nogales, Arizona (1,035), Calexico, California (112), and even San Diego (1,589). In 2001 there were 1,714 auto thefts and 1,656 vehicle burglaries in Chula Vista, representing 44 percent of the city's total crime index.[54]

The report examined the top ten public lots from which vehicles were stolen and found, among many other insights, that experiences varied for two types of lots: those near high schools and colleges, and those near swap meets, trolley stops, department stores, and movie theaters. For the first type, the recovery rates were relatively high (67–75%), suggesting that the theft was for transportation or joyriding. For the second type, which accounted for seven of the ten lots studied, the recovery rates ranged from 37 to 50 percent, indicating that the theft was for export or dismantling for parts. In these cases, thieves targeted the lots where cars would be parked for at least one to three hours. After the theft, a thief could be in Mexico in a matter of minutes.

None of the lots studied had the full arsenal of countermeasures to reduce crime: "electronically armed ticket entry system with staffed exit points for ticket recovery, cameras, active security, and perimeter control," but when one participating mall with a parking lot that literally abuts the border installed "electronic ticketing triggered gate arms, staffed exits to collect tickets, and extensive cameras and security patrols, vehicle crimes dropped to near zero."[55]

Given that the community is adjacent to the entrance into Mexico, authorities also examined whether license-plate cameras at the border would stop the flow of stolen vehicles. In their interviews with border agents, they were told that their primary mission was national security, not auto theft; when the border agents attempted to stop vehicles going into Mexico, massive traffic jams resulted. The best form of intervention took place in lots where auto theft was concentrated.[56]

Ronald V. Clarke listed the following recommended actions in his manuscript "Thefts of and from Cars in Parking Facilities," submitted to the Center for Problem-Oriented Policing in 2002:

1. Hiring parking attendants

2. Improving surveillance at deck and lot entrances/exits

3. Hiring dedicated security patrols

4. Installing and monitoring CCTV

5. Improving lighting

6. Securing the perimeter

7. Installing entrance barriers and electronic access

8. Adopting rating systems for security features

9. Arresting and prosecuting persistent offenders.

Responses that he claimed had limited success included these:

10. Conducting lock-your-car campaigns;

11. Warning offenders;

12. Promoting car alarms and other "bolt-on" security devices;

13. Using decoy vehicles; and

14. Redirecting joyriders' interest in cars.[57]

Given Clarke's list of generally—though not universally—accepted recommendations and the insights derived from research, a convergence of opinion appears to be developing over the strategies that are both most effective and most acceptable for using the built environment to combat auto theft. First is to create natural surveillance, or ways to maximize public visibility. Second is to reinforce the proprietary nature of the space—let the public take "ownership" of it. Third is to clearly differentiate between public and private spaces and control access. These principles seem to be the drivers behind effective measures to prevent crime in general, and auto theft in particular, through environmental design.

A CHANGED PEOPLE, A TRANSFORMED ENVIRONMENT, AND A DROP IN AUTO THEFT

From the days when garage doors were left open all night to living with the specter of CCTV cameras watching our every move in public spaces, we've come a long way as a society. No longer do Americans typically leave their car doors unlocked at night or leave keys in their cars. Our collective insecurities continue to shape the built environment in the race toward the creation of defensible spaces. But as research has shown, our homes, communities, and public areas provide imperfect, though sometimes very effective, means for thwarting auto theft.

The United States will likely never reach the stage where "indiscriminate video surveillance raises the specter of the O[r]wellian State."[58] We have a Bill of Rights that limits the power of the federal government and ensures privacy, although breaches do take place. By contrast, the United Kingdom, which has no such controls and has more video surveillance per capita than any other nation worldwide, illustrates the potential for problems. Because "it is relatively easy to find footage from parking garages, housing developments, department stores, and offices that may have commercial value," breaches can occur: "Cameras may record couples intertwined in office stock rooms, elevators or cars; women undressing in department store changing rooms; or husband and wives engaging in domestic squabbles. Such scenes are sold commercially in UK video stores."[59] Yet crime has, in fact, gone down.

In the United States, which has not resorted to such extremes, crime overall, and auto theft in particular has also decreased dramatically over the past several decades. In some communities, crime prevention through environmental design has been an important addition to the arsenal against auto theft. Continued refinements in field technology, data gathering and interpretation, and use of personnel toward the creation of multilayered approaches to preventing crime suggests that the built environment may become an increasingly vital tool in the arsenal against auto theft. However, one can't help but wonder whether the downward trend is permanent. While there are reasons for optimism, history suggests that nothing, including the incidence of crime, remains the same as time passes.

Car Theft in the Electronic and Digital Age (1970s–Present)

It all boils down to greed. I want that. . . . Drugs made me greedy.

"JOSEPH," A CAR AND MOTORCYCLE THIEF, 2011

Sitting calmly beside Sergeant Steve Witte in the Chula Vista, California, police station, "Joseph," a convicted automobile and motorcycle thief with links to the Mexican cartels, was candid in recounting his past miscues. A high school dropout who started taking drugs at age fourteen and who ultimately made one hundred thousand dollars a year stealing motorcycles, Joseph got his start by stripping cars (Hondas and Acuras) for parts to support his street racing activities. He looked for unmolested cars, daily drivers, "because they were not abused." Ironically, the car that he had invested so much in to race was stolen from his apartment's parking lot, even though it was equipped with two kill switches, a five-hundred-dollar alarm system, and a detachable steering wheel! When authorities found the car and Joseph saw it, he was appalled. He recalled, "I never saw a car stripped the way this one was." Rather than pay a $150 recovery fee, he handed over the title to the police and said goodbye to his once-prized possession. Joseph's recollection was matter-of-fact rather than tinged with sadness. Essentially, he had disconnected himself from the vehicle and the crime; it was almost as if in the end he was not a victim. After all, it was insured.[1]

In a decided fashion, post-1970 film and literature featuring auto theft celebrated rather than decried the act of automobile theft with one caveat—namely, that the car stolen was insured. That qualifier granted, professional car thieves were elevated to an almost heroic status, and amateurs were portrayed in a comic light. Automobiles were often seen as objects of desire, but at the same time in various chase scenes, police cars and everyday "drivers" were depicted as disposable objects, wrecked and destined to be forgotten. In most cases the cars stolen were luscious, high-performance vehicles normally not attainable by any but the highest class of thieves and bandits.

The two *Gone in Sixty Seconds* films—one done in 1974 and the other in 2000—entertained millions of viewers worldwide. These two films celebrated both the professional thief and the high-performance, elegantly styled automobile. And while the story was appropriately centered in car-culture-dominated southern California, the original 1974 version was written by an unlikely outsider who had come from Dunkirk, New York, H. B. "Toby" Halicki (1940–1989).[2] Halicki, with no formal education in film or practical experience in the industry, conceived, wrote, directed, produced, and starred in the version released in 1974. From a rough-and-tumble Polish-American family, Toby began his work life as a tow-truck driver before migrating to California, where he succeeded in various businesses, including automotive recycling, body-shop repairs, and real estate. It comes as no surprise that in an early scene Halicki handles a tow truck, pulling a car like an expert driver behind the wheel of a sports car. He put together a remarkable low-budget independent film, relying on friends, family, everyday police officers, firefighters, and pedestrians to play supporting roles. And, quite modestly, Halicki lists the star of the film as a 1973 Mustang Mach I named Eleanor.

Halicki plays the part of Maindrain Pace, a respected insurance investigator and owner of Chase Research by day. At night and in and around parking lots, streets, the chop shop, and dealerships, however, Pace is the head of a highly organized car-theft ring. Despite what one might think of his illegal activities, Pace is a criminal with principles, for he will not steal a car that is not insured (ironic, given his day job!). As the film opens, the work of a chop shop is detailed: valuable tagged parts, along with the Vehicle Identification Number (VIN) sticker, are "switched over" from a wrecked red Dodge Challenger to one stolen from an airport parking lot. A bit later, with an order from an Argentine general to steal forty-eight cars in four days, Pace and his associates quickly get to work. Here the film illustrates the many ways a professional thief can steal a car without leaving a mark on it. Members of this outfit do not make mistakes; Pace states to one of his associates, "The amateurs are in jail. Professionals never are caught." He goes on to explain that the only professionals who are in jail are those who were sloppy. Each associate is given a specially equipped briefcase containing tools, magnetic license plates, and anything else that one might need to quickly and cleanly "boost" a car. And several of these devices are shown in action—the Slim Jim, a door button pry bar, and a separate ignition switch. Newer additions to the briefcase include a walkie-talkie and a compact key cutter. Several car-culture notables from the era play minor roles in the film—Parnelli Jones, J. C. Agajanian, and Tony Bettenhausen. Given the absence of a script, the flow of the film is rather remarkable, culminating with a thirty-four-minute chase scene and a jump that left Halicki with ten crushed vertebrae and a limp. Among the cars stolen were a 1924 Rolls-Royce Silver Ghost (Eileen), a 1970 Jaguar E-type (Claudia), a 1959 Rolls-Royce Phantom V (Rosie), a 1972 Ferrari Daytona 365 GTB 4 (Sharon), a 1973 Jensen Interceptor (Betty), a 1971 Citroen SM (Patti), a 1962 Ferrari 340 America (Judy), a 1971 Chevrolet Vega (Christy), and a 1967 Lamborghini (Tracy). The connection between the beautiful cars and women's names raises an obvious inference concerning the hot cars and sexuality, although the Chevrolet Vega appears as an outlier. It has been postulated that the Vega was a Cosworth model, but that remains only a conjecture. Despite the low budget, the absence of professional actors, and organizational methods that children have exceeded when making home movies, Halicki may have succeeded in ways in which the 2000 release starring Nicholas Cage and directed by Dominic Sena fell short. While it may be argued that Halicki was far more interested in making a chase movie than one illustrating the nature of auto theft, the most enduring message is the scene that features a battered Eleanor still running at the conclusion of the film. Detroit iron made back in the 1960s was nearly indestructible, if we are to believe Hollywood!

Three years after the release of Halicki's *Gone in Sixty Seconds*, another feature film centering on auto theft appeared, this time a zany comedy with the title *Grand Theft Auto* (*GTA*).[3] A cross between *Gone in Sixty Seconds* and *American Graffiti*, *GTA* was the first film directed by Ron Howard but also drawing on the talents of B-grade film creator Roger Corman, who served as executive producer. The plot is as silly as it can be, interrupted with mindless special-effect explosions and car chase scenes and crashes and filled with exaggerated characters. Despite its flaws, the film made money, parlaying a $600,000 investment into a $16 million profit.

The *GTA* story tells of two young lovers who want to get married, Sam Freeman (Ron Howard) and Paula Powers (Nancy Morgan). Paula, the daughter of wealthy gubernatorial candidate Bigby Powers, is a headstrong and independent young woman, so characteristic of that day, and she is dead set on marrying Sam, contrary to the wishes of her parents. Instead of Sam, Bigby wants his daughter to marry wealthy Collins Hegeworth (Paul Linke), a foolish prig of a young man to say the least. After the young couple elopes from Los Angeles to Las Vegas, taking off from the Powers estate in a 1959 two-tone Rolls-Royce, a fantastic chase ensues. With a reward on their heads and numerous pursuers, they careen through the desert wastes of California and Nevada, ending at the Nevada Speedway. Love triumphs in the end, of course, as the rich young girl chooses to follow her glands rather than her pocketbook.

Beyond the message that young love trumps money, *Grand Theft Auto* contains an innocent view of auto theft. From the moment Paula "borrows" her father's Rolls-Royce to the demolition derby scene at Las Vegas, characters "borrow" the cars of others to suit their immediate convenience. In Paula's view, she is merely "trading with dad," since she cannot drive off in her own Fiat X1/9 because he has taken the keys away from her. Later in the story, Collins, the rejected suitor, drives off in a Dodge Charger and later switches to a battered yellow pickup truck. Collins's mother, concerned for her son, borrows a servant's Volkswagen Beetle. Two gas station mechanics (1977 clones for Beavis and Butthead) borrow a kit car that was in their garage for service and a Chevy Luv pickup with a camper attached. A policeman borrows a bus filled with senior citizens, who do little to protest when they discover that they are going to Sin City. In sum, *Grand Theft Auto* reinforced the notion that stealing a car was not really stealing, just temporarily using a motor vehicle because of a personal need.

The late 1970s were good years for goofy comedies about car theft, such as *Corvette Summer,* released in 1978.[4] Starring Mark Hamill as Kenny Dartley and Annie Potts as Vanessa, *Corvette Summer* is entertaining to a point before its slow pace detracts from the story. The film begins with a high school shop class going to a salvage yard to pick up a project car. Kenny ends up saving a Corvette Stingray from the crusher, and soon he leads the class in what becomes a custom restoration of the shark-nosed vehicle. Kenny sees the final product as "perfect"; but his shop teacher, Mr. McGrath, cautions his student not to get too involved with the car, because automobiles "always let you down." The viewer can see that in reality, the restored car is one of the most garish creations in all of automobile history!

During a night of celebration at a local cruise-in, the car is stolen by organized car thieves. While authorities are resigned to the loss, young Kenny refuses to give up. He begins to search for the vehicle, and his odyssey eventually takes him to Las Vegas. On the road, Kenny meets a spindly-built young woman who becomes the love of his life—Vanessa, a wannabe hooker who drives a customized lovemaking van. At first, Kenny resists Vanessa's advances, preferring the love of his lost car to that of a young woman who was excessively thin before thin was "in." Indeed, his love is for a car that is not his, but rather is owned by his high school, and thus his obsession is with his work and creation, rather than with personal property.

After arriving in Las Vegas and sighting the car several times, Kenny finally tracks it down to the Silverado Auto Body Shop, where he discovers not only the car but the fact that his mentor back in Los Angeles, Mr. McGrath, is a part of the stolen-car ring. Thus this film shows an adult at the root of auto-theft problems that involve a juvenile, a common theme going back to the 1950s (see chapter 2). After accepting the loss of the car and actually taking a job at the body shop, Kenny decides to do the right thing and take the car back to Los Angeles, even though he has heard arguments that "a car is a commodity to be bought and sold," that "the crime was victimless," and that the insurance company can afford to pay for it. Finally, after a chase scene in the desert outside Las Vegas and the return of the car to the high school, Kenny comes to his senses and walks off with Vanessa. In the end, the girl does become more important than the car; a human relationship triumphs over one with a machine.

The girl is also far more important than the machine in *Breathless,* the 1983 remake of Jean-Luc Godard's nouvelle vague classic *A bout de Souffle.*[5] *Breathless* starred a young Richard Gere as drifter Jesse Lujack, who after a brief Las Vegas fling becomes obsessed with Monica, a UCLA exchange student from France, played by Valerie Kaprisky. As the film begins, Jesse, a high-energy punk "who rolls the dice too much," is using a screwdriver and two blades to steal a Porsche 356A coupe in front of a Las Vegas casino. He comes alive while driving and listening to Jerry Lee Lewis ("You Leave Me Breathless"), but his trip to Los Angeles is interrupted by a police stop and an unintentional murder. On the run and determined to take Monica with him to Mexico, Jesse steals a succession of vehicles while eluding police; stolen are a pink MGB, a 1957 Thunderbird, a Ghetto Buick, an old truck, and a 1959 Cadillac Eldorado convertible. In every case, as Jesse gets behind the wheel, he gains energy, and mobility brings with it self-realization. But as a frustrated male who models his psyche after the comic book character the Silver Surfer, Jesse is shallow, with no long-term future and projecting a tragic end to the adventure. Yet, at times he imagines himself as the Silver Surfer, who once was an ordinary man from a distant planet. Forced to serve Galactus, a planet-eating god, the comic book hero preserves his world and the woman he loves from destruction. Afterward, endowed with powers of infinite movement by Galactus, he searches the galaxy for planets to feed his master, until his encounter with Earth forces him to betray Galactus and thereby redeem himself. But in the end both the Silver Surfer and Jesse are left trapped on the planet he saved. Forced into a decision to either surrender or pick up a gun and die at the hands of police, he chooses the latter, perhaps realizing that Monica and Mexico are forever out of reach.

SONG AND CAR THEFT

Unlike film, song has rarely exploited the topic of automobile theft. Its conspicuous absence is ironic, given that particularly since World

War II the car has been at the center of popular music.[6] Yet, on the few occasions when the act of stealing cars has been featured in lyrics, themes common to those used in films emerge, but with a far more desperate and dark tone. Car theft as connected to adventure and sexuality stands out in Joe Bonamassa's "Tennessee Plates" (2011) and Sting's "Stolen Car" (2003), but then so do failure, broken relationships, and loneliness. These endings are little different from that experienced by Jesse Lujack in *Breathless.*

The psychological highs of reckless abandon gained by illegal mobility, however, eventually lead to dire straits. Employing a story not terribly different from that of *Bonnie and Clyde,* "Tennessee Plates" (first performed by Randy Travis in 1998) tells of a girl "shivering in the dark" on a cold night and how that scene begins a tale of bank robberies, car thefts, an exhilarating ride crossing "the Mississippi like an oil slick fire," and a man left for dead on the interstate. Yet the trip ends in confusion, for the hero wakes up in a hotel room "in original sin," with no answers to his current dilemma.[7]

Of all the songs featuring car theft, perhaps the one best known—and the one with the deepest psychological overtones—is Sting's "Stolen Car" (2003). About a poor boy who hotwires a rich man's car, the song's lyrics, coupled with its rhythm, reach deep inside the listener, evoking a sensual experience that involves adultery, the lingering smell of cologne, and the presence of automobile lights in the midst of darkness:

> Late at night in summer heat. Expensive car, empty street
> There's a wire in my jacket. This is my trade
> It only takes a moment, don't be afraid
> I can hotwire an ignition like some kind of star
> I'm just a poor boy in a rich man's car
> So I whisper to the engine, flick on the lights
> And we drive into the night[8]

Bruce Springsteen's song "Stolen Car" (1980) also possesses a dark angst, perhaps connected to adultery. The common everyman protagonist agonizes over a marriage gone bad, riding in a stolen car during a "pitch black" night, horribly fearful and alone, desperate and at the end of his line. Yet, as Springsteen laments, "Each night I want to get caught. But I never do."[9] Similarly, but in the context of a very different urban environment, the Beastie Boys indicate their own take on the matter with their hip-hop-genre title "Car Thief" (1989). In this song too, the main character's life is coming apart "at the seams" as a consequence of violence, plenty of drugs, and a disconnected urban lifestyle that has resulted in human worthlessness. He ends up incarcerated at the "Mountain," while his former friend cuckolds his girl.[10]

FICTION CONFRONTS FACT

Despite cultural representations suggesting otherwise, auto theft as it took place during the recent past was more often than not a deadly serious activity. That fact was forcefully illustrated during Senate testimony given in 1979, when a hooded witness referred to as "John Smith" appeared before the Permanent Subcommittee on Investigations of the Committee on Governmental Affairs. Smith was serving a five-year sentence for conspiracy to transport stolen vehicles through interstate commerce, and he freely admitted that he had been a member of a ring of forty-five individuals operating in nine states and Mexico. The witness personally had been responsible for fifteen hundred auto thefts; some members of the ring had cleared two hundred thousand dollars a year from its operations. Further, Smith echoed an old theme concerning deterrence: "I have never encountered an automobile locking system that I could not defeat in a very few minutes. I probably never could have gotten into illegal rebuilding if it hadn't been so easy to change the few vehicle identification numbers now on cars and trucks."[11] Smith went on to outline the techniques and methods of the professional auto thief of the 1970s, including details of how he acquired the rosebud rivets that attached VIN plates to the vehicle; the location of "confidential" VIN numbers that manufacturers stamped on frames, marked on inner fenders, and spread on small pieces of paper throughout the vehicles; and how a welding tool could erase motor numbers. Yet, eventually, Smith's attempts to obliterate a motor number were detected by an infrared camera developed by the National Automobile Theft Bureau.

According to Smith, however, key copying proved to be the least destructive and most effective technique to steal a car. He preferred General Motors cars from the 1960s, for these models had the same door and ignition keys. Once the door lock was removed and a dummy put in its place, Smith said he would walk "off to my car, set down and tear the lock apart and read the tumblers in it and cut a key for it. It would take about 30 seconds."[12] To counter the ease with which he did his work, Smith stated that all manufacturers had to do was to make their lock components out of harder metal.

Key cutting, Smith said, was the "blue collar" approach to auto theft. He also had a "white collar technique." After finding the desired car, he would ascertain from a license plate holder or dealer sticker where the car was purchased. The next step was to use the license plate number to contact the department of motor vehicles, which, without checking the identity of the caller, would furnish the name and address of the owner. Smith would then wait until about 4:30 p.m., a time when car dealership staff were always at their busiest. Posing as a locksmith, the would-be thief would state that a woman shopper had lost her keys at the mall. Hurried, the dealer's employee, without questioning, gave the "locksmith" the key code. Smith then had all the necessary information, and using tools purchased from a business that sells repossessor supplies, he was able to complete an unnoticed theft in quite a white-collar style.

Smith's testimony was important in shedding light on the activities of the car thief. However, a major aim of the hearings was to uncover information related to criminal organizations of the day, particularly the inner workings of the chop shop. Alex Jaroszewski, who at one time had been a midlevel operative in an extensive Chicago chop-shop operation with links to organized crime and in November 1979 found himself in the federal witness protection program, gave valuable testimony on the subject. He had been gradually brought into the ring in the 1970s by salvage yard owner Steve Ostrowsky. Over time, Jaroszewski was increasingly given more responsibilities and began to acquire knowledge of the top end of the business, including some of the illegal activities headed by Jimmy "the Bomber" Catuara. But first

Jaroszewski had to get the tools, techniques, shop, and connections with Ostrowsky's south Chicago salvage yard, where cars were stolen to order and parts were selectively taken off the hot cars and then distributed in an extensive salvage-yard network that was connected by private phone lines throughout the Midwest. Jaroszewski later recalled how chopping was done:

> First, he [a mechanic] unbolted the front end, which is the entire front section of the car including the fenders and the hood. This is the most valuable part of the car because it is the section most frequently damaged in accidents. The windshield was then cut out and the doors unbolted and removed. Next, the seats were removed. Then, with a torch, he cut the posts which connect the roof to the dashboard and cut through the floor width-wise in the front seat area, enabling him to remove the entire cowl. . . . This left him with the roof section of the car connected to the back end which is often referred to in the salvage business as a rear clip.[13]

Once the car was dismantled, the usable parts were taken to South Chicago Auto Parts, while the marked frame, engine, transmission, and other components were carted to a crusher to be recycled as scrap iron. It was a profitable activity, as these stolen parts were sold at one-quarter or less of the price of parts purchased from manufacturers, and availability was almost immediate. The business, however, while on the surface rather tame, had a violent undercurrent. As Jaroszewski discovered one morning at breakfast, not only was his boss an enforcer for the Mob, but so was his boss's boss, a thug by the name of Billy Dauber, another salvage-yard owner. With Chicago police on the take, arrests occurred only rarely, and then release was quick; the police even returned tools to the thief! But the real danger that these chop-shop employees faced was from other gang members. In time Ostrowsky, his partner Holzer, shop operator Timmy O'Brien, the big boss Catuara, and enforcers Richie Ferraro and Richard Pronger were all murdered in a territorial shakeup. For a time afterward, Billy Dauber emerged as second in command to Albert Tocco in a new chop-shop structure in Chicago.[14]

CHOP SHOPS AND THE FEDERAL GOVERNMENT RESPONSE

By 1982 the federal government had stepped up its efforts to reduce auto thefts, particularly those by the organized crime sector, by taking direct aim at chop shops. The House Subcommittee on Consumer Protection advanced a bill that mandated the marking of critical car parts, and the Senate concurred with a companion bill. Although Congress was behind these measures, manufacturers and the Reagan administration were not. In fact, during a hearing on the bill, one administration official spoke in favor of the bill and another testified against it. Diane K. Steed, deputy administrator of the National Highway Traffic Safety Administration (NHTSA), presented the administration's position at the hearing. She said, "The industry is already facing a serious cash-flow problem . . . and $100 million worth of regulatory burdens on the motor vehicle industry is not in keeping with the regulatory relief effort."[15]

Congress ultimately prevailed, and the 1984 Motor Vehicle Theft Law Enforcement Act was made into law. The Theft Act was designed to reduce the incidence of motor vehicle thefts and simplify the tracing and recovery of parts from stolen vehicles. The act directed the secretary of transportation to issue a theft-prevention standard requiring manufacturers to inscribe or affix numbers or symbols on major parts of high-theft passenger cars for identification purposes. The legislation also addressed issues such as criminal penalties, the export of stolen motor vehicles, and comprehensive insurance premiums.

In October 1985, the Department of Transportation implemented the Federal Motor Vehicle Theft Prevention Standard, which required manufacturers of designated vehicles to inscribe or affix the VIN onto the following major parts: engines, transmissions, fenders, doors, bumpers, quarter panels, hoods, and deck lids or tailgates and/ or hatchbacks. In the case of engines and transmissions, either the seventeen-digit VIN or an eight-digit VIN derivative had to be engraved or stamped. Manufacturers could meet the affixation requirements with indelibly marked labels that could not be removed without being torn or rendering the number on the label illegible. The labels also had to leave a residue on the part after being removed.

As a further theft deterrent, the 1984 act allowed for an exemption from the parts-marking requirements for certain brands where anti-theft devices were installed as standard equipment in factory-delivered cars. The act limited each manufacturer to exemptions on two models per year. The manufacturer had to petition NHTSA for an exemption, which would be granted if the devices were determined to be as effective in reducing and deterring motor-vehicle theft as compliance with parts-marking. The common features of antitheft devices installed as standard equipment for which exemptions were granted included "passive" systems, those that engaged automatically without any extra action by motorists. Such systems were activated automatically by removing the key from the ignition and locking the doors. Their sensors, located in the doors, the hood, the trunk, and the key cylinders, activated an alarm after an unauthorized entry attempt. Approved systems also had a starter or ignition interrupt and power (battery) protection. Most systems that were granted full exemptions featured an audio or visual alarm or both, and some of the General Motors systems used a passkey. Systems that were granted a partial exemption had marked engines and transmissions.

After Congress acquired more information, the Anti–Car Theft Act of 1992 was passed. This legislation built on the 1984 act in several ways: Federal penalties for auto theft were enhanced; a grant program was authorized to help state and local law enforcement agencies concerned with auto theft; and experts were called on to look into and report on motor-vehicle titling, registration, and salvage. In 1994 the National Motor Vehicle Title Information System was established, and states were required to participate. Under this system, the Theft Prevention Standard was expanded, rules were established to check whether salvage or junk vehicles had been stolen, and the U.S. attorney general was charged with maintaining a National Stolen Auto Part Information System. Selling or distributing stolen marked parts became a federal crime.

The significance of the chop shop, a place where "every car gets a new life," in organized auto thievery is highlighted in Peter Werner's 1987 film *No Man's Land*. Werner's penchant for the subject of auto theft had resurfaced a decade after the release of *GTA*, but with a far more violent and serious twist.[16] In *No Man's Land*, skilled thieves steal

1980s Porsches with only Slim Jims, slide hammers, and some force to break a steering wheel lock. Aided by insider collaboration from both the police and "a guy at the DMV," these thieves thrive at the expense of the wealthy, some of whom no doubt live the "life style of the rich and aimless." There seems to be no sympathy expressed for the car owner in this film, and perhaps its greatest virtue is various scenes in which Porsche 911s are driven hard on the street. The real danger in committing this kind of crime was not from the police or the actual act of theft, but from rival rings and contested turf.

Starring Charlie Sheen (Ted Varrick) and D. B. Sweeney (Benjy Taylor), *No Man's Land* is a tale of luscious cars, camaraderie among thieves, and one beautiful and rather innocent young woman. Sweeney is cast as an undercover cop, while Sheen plays the role of a superficial, cocky rich kid turned master auto thief and ring leader. Rookie cop Benjy takes on the identity of mechanic Bill Ayles and gets a job at Technique Porsche, a shop that is suspected of being a place where stolen Porsche 911s are cloned and chopped. The film opens with a scene from this shop in which a Porsche 911 VIN is being cut from the frame of the car, while its speedometer and steering wheel are being removed. (The parts shown actually came from a Porsche 356, whereas the car shown is a 1980s Porsche 911 Targa.) As the plot evolves, mechanic Bill develops a friendship not only with master thief Ted Varrick, but also with Varrick's gorgeous sister. Gaining Varrick's confidence, Ayres teams up with him to steal numerous Porsches, but their success leads them into a conflict with a rival gang. As the undercover cop gets deeper and deeper into the ring, where does his allegiance lie? In the end the violence turns deadly, the romance goes sour, and Ted is shot to death by his former friend Bill. One wonders whether the story told in *No Man's Land* is Hollywood hyperbole, perhaps not at all representative of the crimes actually committed as reflected in statistical summaries.

STOLEN CARS AND STATISTICS

As the decade of the 1980s opened, *New York Times* writer Charles Klaveness remarked, "My car used to be a 1974 Ford Bronco, but now it's a statistic. Last year, 110,881 motor vehicles were reported stolen in New York State, according to state and federal crime reports. No figure is available on what it cost insurance companies to settle resulting claims, but it was more than ever before—the National Automobile Theft Bureau says the average value of a stolen vehicle has gone up 60 percent in the last five years, and it wasn't going down before."[17]

Initially, the surge in auto theft in the late 1970s and early 1980s was blamed on juveniles. And while recovery rate statistics certainly suggest that youthful offenders were often responsible for the thefts, by 1984 a perceptible transition had occurred in which organized thieves emerged as the focus of law enforcement efforts. On the local level, this shift was reflected in the rhetoric of New Jersey state senator Donald T. DiFrancesco, a champion of anti-auto-theft legislation. In November 1980, DiFrancesco, representing a constituency that was becoming increasingly impatient with the inability of authorities to curtail neighborhood crime, argued that youthful perpetrators should be treated with more "maturity." In an op-ed essay published in the *New York Times,* DiFrancesco asserted, "In 1978, persons under the age of 18 accounted for 60 percent of all arrests for breaking and entering, automobile theft and arson, and for 40 percent of all arrests for robbery. . . . Juvenile violence and criminality are no longer a rising tide; it is a terrifying and destructive wave that, if not checked, will engulf urban, suburban and rural New Jersey."[18] Two years later, DiFrancesco introduced legislation "to give the police and prosecutors additional tools in catching professional car thieves and sending them to jail." The tools, he added, would "include improved cooperation between law enforcement officials and insurance companies by making the companies immune from prosecution by providing the police with information about auto theft claims." A final measure in DiFrancesco's legislative package was designed "to prevent auto-theft cases from being thrown out of court because the owner of the car fails to attend the trial. It would permit a certificate of title to be entered into court as proof of ownership, sparing the victim from being required to testify."[19] The shift in DiFrancesco's legislative efforts reflects a subtle but important transition in perceptions about auto theft. After 1980 law enforcement, the insurance industry, and inventors of antitheft devices clearly targeted organized crime. Minors were still an important part of the story, however, because they often worked as low-level operatives within complex hierarchical organizations.

Despite dire reports of the 1980s and 1990s, when it comes to auto theft in America, this period can ultimately be characterized as one when technology and organization, harnessed together, effectively put the brakes on stealing cars in America. That is not to say that the problem was solved, by any means, or that a major cultural shift has taken place with regard to the casual nature of the act. However, every statistical and quantitative indicator reflects a victory of sorts in the war against this crime; it has come as a consequence of introducing sophisticated electronic technology coupled with intelligent policing. Yet the story is far from over: the same technology that has played such a large role in shutting down the joyrider, or a slight variant, can also be used by a new type of thief, akin psychologically to a computer hacker. And indeed, computer gaming has culturally reinforced a lawless, anti-authoritarian value system that is adhered to by many people living on the margins of American life (see chapter 6 for additional details).

In 1986 there were 1,224,127 motor vehicle thefts in the United States. A decade later, that number had peaked at 1,395,192. Yet, by 2010, the total had dropped precipitously to 737,142, or 39.8 percent less than in 1986 and 47.2 percent less than in 1996.[20] Similar trends can be extended, with few exceptions, from national aggregate data to states and cities. For example, in Los Angeles total motor-vehicle thefts dropped from 57,331 in 1988 to 18,391 in 2009.[21] These lower numbers have been coupled with higher recovery rates, painting an even rosier picture concerning the recent past. Of course, hot spots—states, cities, and neighborhoods—are still a major concern for law enforcement officials, and because there is spatial movement at the local level, the crime continues to confound police, the insurance industry, and car owners. The statistics indicating a reduced threat across the nation are of little consolation to those living in the geographical areas that are the worst for car theft in America.

As might be expected, the state of California reported the greatest number of auto thefts in 2010. The next-largest numbers of thefts occurred in states that border Canada or Mexico or face the vast European, Asian, and Latin American markets (see table 4.1). California stands out in that its number of auto thefts in 2010 is more than double that of the next state, Texas, and greater than the combined numbers of thefts in the next three states (Texas, Florida, and Georgia). Conspicuously absent from this list are states that have few large urban centers, those with primarily agricultural economies, and those with lower population densities.

Large urban areas also differ significantly in auto-theft statistics. The metropolitan statistical areas (MSAs) with the greatest number of auto thefts reported in the decade from 2000 to 2010 are listed in table 4.2 for 2000 and in table 4.3 for 2010. By 2010 the Los Angeles MSA had become the auto-theft capital of the United States and was followed by Chicago, New York, Houston, San Francisco, Dallas, Detroit, Atlanta, Miami, Riverside, Seattle, Washington, D.C., San Diego, Phoenix, and Philadelphia.[22] The critical characteristics of these cities are their large pools of autos, their proximity to the interstate highway system, their access to a significant homeland or international market reachable by land or sea, and the existence of a sophisticated network of people involved in auto theft.

Yet another development is the rate of auto theft measured as the number of cars stolen per one hundred thousand persons. In 2000 the three cities with the highest rates were large centers—Phoenix, Miami, and Detroit (see table 4.4)—where the number of thefts was also significant. But surprisingly, in 2010, the largest cities, including Los Angeles, Chicago, and New York, did not even make the list. Smaller cities such as Fresno, CA; Modesto, CA; and Bakersfield-Delano, CA, outranked the major centers (see table 4.5). Although the dramatic changes in ranking can be attributed to changes in the size of the base relative to the number of thefts (e.g., the Los Angeles MSA was redefined to incorporate a larger area), one interpretation suggests the growth of a vibrant shadow economy involving these smaller communities. "The common thread among U.S. cities with the highest car theft rates [in 2010] is that many of these areas are actually border cities and towns that have become a pipeline for car thieves smuggling everything from weapons and money to drugs through Mexico. Simply put, stolen vehicles in most border towns are used to commit other crimes. Yakima, Washington . . . cites rampant criminal activity such as drug dealing and heavy gang activity as the cause of most car thefts. Yakima also happens to sit on a main road that runs between Canada and Mexico."[23] (See chapter 5.) The new landscape of crime defined by cities and regions linked through organized crime takes on added significance when viewed against the drop in auto theft occurring nationwide: an effect of that overall decrease was to increase the geographical concentration of the crime. See table 4.6 for a list of urban neighborhoods that illustrates the existence of geographical "micro-bursts," or highly concentrated areas where cars are frequently stolen.

The statistics also reflect differences in the typical perpetrator of auto theft. As we have seen, research suggests that when cars were scarce during the 1930s, and again during World War II, most cars were stolen by older, more experienced professionals, but after the 1930s and in the postwar era, when autos were readily available, young people predominated.[24] Now, once again, car thieves are mostly adults.

Because California is prominent in auto thefts nationwide, data on arrests for motor-vehicle thefts by adults and juveniles in that state provide a vantage point from which to view trends. The period from 1999 to 2010 provides useful insights and displays a clear pattern: adult arrests increased (from 66.9% to 80.2%) while juvenile arrests declined (from 33.1% to 19.8%). During the same period, the number of auto-theft arrests skyrocketed from 1999 to 2004 (from 19,728 to 30,731), then dropped precipitously by 2010 (13,091), mirroring national trends. Table 4.7 compares the auto-theft arrest totals for adults and juveniles during the 1999–2010 period. As shown in table 4.8 for the years 2004 and 2010, within the shrinking pool of arrestees, the age group thirty to thirty-nine showed growth as a percentage of the total, while arrests in the twenty-nine and underage group declined. Historically, car theft has been largely a male crime and continues to be such. As evidenced in California, approximately 80 percent of those arrested for auto theft between 2004 and 2010 were males, although the data suggests a subtle growth among women (see table 4.9).

CULTURAL REPRESENTATIONS OF VIOLENCE IN THE CITY

A prime example highlighting the connections between violence in an urban environment, race, and auto theft can be found in the 1993 release *Menace II Society* and Spike Lee's 1994 *New Jersey Drive*.[25] Without doubt, these two films are by far the most emotionally powerful and realistic of all twentieth-century films that focus on the topic of auto theft. *Menace II Society* paints an image of what it was like to live in Los Angeles's Watts neighborhood during the 1990s, including carjacking, auto theft, and drive-by shootings. *New Jersey Drive,* as it opens, introduces viewers to the central figure in the story, Newark car thief Jason Petty (Sharron Corley), one of a large group of aimless young African Americans who steal cars, in the process making Newark, New Jersey, the car-theft capital of America. They do it to "put on a show"; it "didn't matter what the car" was. As it turns out, a struggle between young car thieves and the Newark police escalates into a deadly war. The auto-theft squad headed by Lieutenant Emil Roscoe (Saul Stein) brutalizes the young thieves at every opportunity, and the excessive force only exacerbates a volatile situation. The power of this movie goes beyond simply the characterization of auto theft in a major American city; a complex picture of hopelessness, despair, social disintegration, racism, hate, and comradeship quickly emerges.

Lighter, and whiter, the 2000 *Gone in Sixty Seconds* contains plenty of juvenile humor, but it is set in a time when several contemporary truths concerning auto theft surfaced.[26] Unlike Toby Halicki's 1974 original, this remake contains significant elements of violence, perhaps even violence for violence's sake. The most significant theme of the film contrasts the practical wisdom of the organized professional with the impulsiveness of the young amateur. At the beginning of the film, a gang of young thieves led by Kip Raines (Giovanni Ribisi) steals a Porsche 911 by throwing a brick through a dealer's showroom window, opening the key box, and crashing through the showroom glass. Somewhat miraculously, not a scratch appears on the stolen car. A foolish flirtation puts the trio in the sights of the law, and they are followed back to their hideout, where they elude capture but also lose the cars they have stolen. Afterward, these young men are described by a seasoned professional as "little boys in nursery school." And while the inexperienced young men play a part in the redemptive boosts that follow, lapses in judgment ultimately reduce them to having their "decision-making processes taken away from them." Kip's failure leads to the recruitment of his retired expert car-thief brother, Memphis Raines (Nicholas Cage), who, along with former collaborators including Otto (James Duval), is forced by extortion to steal fifty high-end vehicles in four days, with South America as the ultimate destination. These retired pros are now doing things like teaching kids karting and how to restore cars rather than chopping them, and teaching Asian women how to drive. Perhaps their routine lives are a reflection of more difficult times for the car thief during the last two decades of the twentieth century, the consequence of new deterrent technologies and enforcement procedures.

As these cars are boosted, the viewer learns of a host of high- and low-tech techniques used by organized professional car thieves, varying from Slim Jims and slide hammers to computers and electronic frequency detectors. One recent technology introduced by manufacturers deters even the best of these thieves—laser-cut keys featured on new Mercedes. Laser keys, by the way, are not cut by a laser, but by a high-speed titanium bit. It consequently takes the clever ruse of stealing the cars from a police impound to get around this seemingly insurmountable technological barrier. Yet by 2010, defeating a laser cut key is as simple as watching a YouTube video and owning a blank, clay, calipers, a key-cutting apparatus, and a Dremel tool![27]

All ends well by the film's conclusion, as Memphis finally makes peace with his "unicorn," a gold 1967 Shelby Mustang GT 500. Cage, who attended several driving schools and did his own driving stunts in the film, is featured in one of the most memorable of all chase scenes toward the end of the film. And his former lover and girlfriend, female auto thief Sara "Sway" Wayland (Angelina Jolie), adds a feminine touch to what is generally recognized as a masculine criminal activity. It is curious to note that the targeted cars are given women's names— supposedly as code words for the thieves who communicate via radio. Among the fifty "ladies" are the following: a 1962 Aston Martin (Barbara); a 1964 Bentley Continental (Alma); a 1953 Corvette (Pamela); a 1969 Dodge Daytona (Vanessa); a 1957 Ford Thunderbird (Susan); a 1957 Mercedes-Benz 300 SL Gullwing (Dorothy); a 1950 Mercury Custom (Gabriella); a 1961 Porsche Speedster (Natalie); and of course the film's star, the 1967 Shelby Mustang GT 500 (Eleanor).

Despite the superficiality of *Gone in Sixty Seconds,* the film contains some deeper insights, especially concerning the motives of a professional car thief. Sway states while discussing her life working two honest jobs, "I have discovered you have to work twice as hard when it's honest." But stealing cars is more than just making easy money. Memphis explains it this way: "I did it for the cars . . . begging to be plucked. I'd blast to Palm Springs, instantly feeling better about myself." And Memphis tells his younger brother who lacks these sensibilities, "[When you're driving,] the car is you, you are the car."

Chop Shop, a 2007 independent film co-written, edited, and directed by Ramin Bahrani, depicts a very different view of contemporary auto thievery, this time from the vantage point of a twelve-year-old street orphan and accomplice.[28] Alejandro, or Ale, played by Alejandro Polanco, and his sixteen-year-old sister Isamar (Isamar Gonzales), live in the office of a Queens, New York, body shop, eking out a living any way they can. During the day, Ale helps around the shop, putting on side mirrors, buffing hoods, sanding and repairing bumpers, directing traffic into the shop, sweeping up, and locking up at the end of the day. Always hustling, always looking for money, he scavenges for parts or steals hubcaps from an airport parking lot when he takes breaks from his regular tasks. At night, he helps chop a stolen car, dismantling a trunk lid and then carrying it away. In his harsh world of abuse and survival, Ale is nothing more than a victim in a society that has little compassion. At the bottom of the car-theft social pyramid, children like Ale do what they have to do to survive the best they can, and they are exploited at every turn.

INNOVATIVE TECHNOLOGICAL MEASURES

Efforts to correlate stolen-car statistics and violent crime with technological antitheft innovations are fraught with difficulty because of the many confounding and complex factors involved. Since important changes were implemented at the national level after 1980, including antitheft technology and manufacturing, as well as insurance, informational, surveillance, and legislative measures, attempting to isolate the impact of any one of these new strategies would be foolhardy. To be sure, by the 1980s, alarm systems and buzzers reminded drivers to remove their keys, and despite the annoyance they caused, they did reduce joyriding. Also, manufacturers marked vehicle identification numbers on key auto parts of numerous high-theft models, and that stifled some chop-shop operations.[29] Other innovative technological systems that made an impact during the 1980s included the Club, the LoJack, and RFID (Radio Frequency Identification) immobilizers.

Sergeant Steve Witte of the Chula Vista, California, Police Department and head of the San Diego County Regional Auto Theft Task Force, swears by the Club, so much so that he carries some Clubs in the back of his vehicle and sells them at cost.[30] A stout, two-piece adjustable steel bar fitted with a strong lock, the Club is normally attached across a car's steering wheel, making it impossible to turn the wheel a full 360 degrees. In a world of high-tech antitheft devices, it seems a bit counterintuitive that something as low-tech as the Club is touted by anticrime experts. Yet engineers involved in the development of auto-theft countermeasures are the first to agree that all electronic devices, no matter how sophisticated, are vulnerable—albeit only to a select group of thieves with considerable technical skills.[31] But the inexpensive Club often gets the job done by encouraging the would-be thief to move on to an easier target. In doing so, the Club does prevent a crime, even though it may just shift the theft to another car that is perhaps easier to steal but less desirable.

Businessman Jim Winner is usually credited with introducing the Club, although the story is far more complex than the one told on the Winner International website. It was said that Winner, a salesman who had at one time or another hawked vacuum cleaners, chemicals, and pianos, among other things, was already working with mechanic Charles Johnson on an antitheft device when his Cadillac was stolen.[32] Thinking back to his days in the army, Winner recalled the chains soldiers placed on Jeep steering wheels in Korea, and an idea came to him for something that would work as effectively. Johnson's patent describes a device that was easy to use, adjustable to fit a wide variety of steering wheel diameters, and difficult to pry off. Although it was as much a psychological theft deterrent as a physical impediment that significantly kept the steering wheel from turning, it was made of difficult-to-cut case-hardened steel. On early models, thieves could freeze the mechanism with freon and pop the locks off the assembly. But in reality, how many thieves actually carried a can of freon with them? And as a last resort, the five-pound bar could be used as a baseball bat against a would-be criminal in the event of a physical confrontation! As a final pitch, Winner International promised that if a car equipped with the Club was stolen, the company would reimburse the owner of the car up to fifteen hundred dollars to cover the deductible on the owner's insurance policy. In 1991, 147 owners made such a claim.[33]

Technologically, the Club had an arrangement that was similar to the Krook-Lok, sold by mail-order parts supplier J. C. Whitney.[34] The Krook-Lok was a ratcheted chrome steel bar with two hooks that connected the brake pedal of a vehicle to the steering wheel; it was the result of Kitty Zaidener's "Anti-Theft Device for Road Vehicles" 1966 patent.[35] Both the Club and the Krook-Lok had antecedents that went back to the early 1920s: L. S. Baker and W. A. Tabor's Automobile Lock, which connected a levered emergency brake to a foot pedal, and F. M. Furber's invention that bolted to the steering wheel.[36]

After 1987, the Club went through a series of improvements in its locking mechanism. Winner, ever the salesman, positioned his product by offering several less-expensive and more-expensive models than the original bright red device, including a "Designer Club" that came in four neon colors.[37]

The LoJack, very different from the low-tech Club, emerged commercially during the 1980s as perhaps the most successful high-tech antitheft device, a reputation that it enjoys to this day.[38] The origins of the LoJack can be traced to the efforts of former naval aviator and businessman William R. Reagan. A former executive at the AVCO Corporation and owner of an investment banking firm in Boston, Reagan served as a Medford, Massachusetts, part-time police commissioner and was acutely aware of the drug and auto-theft crime problem that was plaguing the United States during the 1970s. He also realized that time was of the essence in preventing a professional car thief from getting to the chop shop.

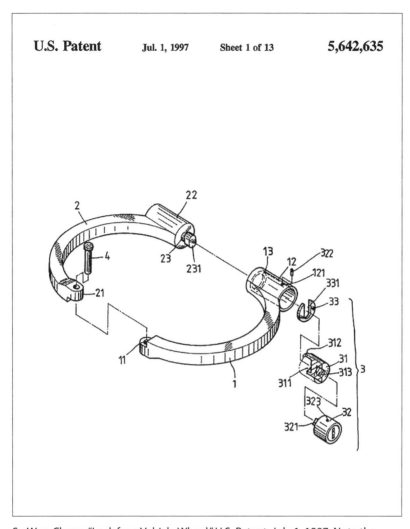

U.S. Patent   Jul. 1, 1997   Sheet 1 of 13   **5,642,635**

Su Wen-Chyun, "Lock for a Vehicle Wheel," U.S. Patent, July 1, 1997. Note the similarity in design to E. E. Chapman's "Vehicle Shackle," patented in 1920. See figure on p. 13.

As the story goes, one morning at 3:00 a.m. Bill Reagan was eating a cookie and drinking milk when he hit upon the idea of hiding a small, independently powered homing device in a vehicle, thus enabling the car to be tracked if stolen. Between 1969 and 1978, Reagan and his co-workers designed and tested a technological system employing transmitters and transponders that was patented as the LoJack in 1979.[39]

Without doubt, transponder technology has done much to reduce auto-theft incidents during the past twenty years. Transponders (transmitters-responders) were first developed during World War II in response to the need to identify friendly and enemy aircraft by radar. A "box" or transmitter, either passive (reflective) or active (in broadcast mode), is placed in the aircraft (or the vehicle). After it "wakes up," it issues a signal; both systems are used for keyless entry and vehicle immobilization.

With a name that evoked the antithesis of "hijack," the LoJack's function was not to prevent an auto theft but to locate the car as soon as the theft was noticed. The owner is responsible for notifying the police. Until this happens, the LoJack does nothing—clearly a real weakness of this technology. Once the police are notified, however, a law enforcement computer starts up a Sector Activation System containing specific vehicle codes that energize the stolen car's box. With the unit now transmitting, a silent code-tracked signal is sent to a police tracking computer installed in a police cruiser, which then displays the stolen vehicle's make, year, and distance and also displays a directional signal.

In 1986 the Motorola Corporation began manufacturing LoJack units for Reagan's firm, and Motorola continued that relationship into the twenty-first century. Beginning in Massachusetts, the LoJack was made available in high-density, high-crime urban areas. In 1989 the Federal Communications Commission allocated a police radio band of 173.075 MHz for stolen vehicles, enabling the corporation to expand nationally to Florida, California, Michigan, Illinois, and New Jersey. By the mid-1990s, the technology was licensed overseas, first in six European countries and then in South America and Hong Kong.

For various reasons, the presence of the LoJack makes auto theft riskier and less profitable and leads to a reduction in the number of such crimes. First and most importantly, the LoJack disrupts the operation of chop shops. Without the device, identifying chop shops requires time-consuming, resource-intensive sting operations. With the LoJack, police following the radio signal are led directly to the chop shop. In Los Angeles alone, the LoJack has resulted in the breakup of numerous chop shops. Second, data collected in California suggest that the arrest rate for stolen vehicles equipped with the LoJack is three times as great as for cars without it (30% versus 10%). Since most thieves are repeat offenders, arrests that lead to incarceration may also provide social benefits via reductions in victimizations while the criminal is behind bars.

On one hand, there is strong empirical evidence that the LoJack reduces auto theft. According to one educated estimate, one auto theft is eliminated annually for every three LoJacks installed in central cities. On the other hand, while Ian Ayres and Steven D. Levitt argued that there was little evidence that the reductions in central-city auto thefts are simply being displaced either geographically or to other crime areas, more recent studies suggest otherwise.[40] In a 2008 unpublished seminar paper at Princeton University, economist Marco Gonzalez-Navarro described a quantitative study examining various regions in Mexico. Gonzalez-Navarro demonstrated that negative externalities existed if thieves could distinguish between LoJack- and non-LoJack-equipped cars, concluding that "most averted thefts were replaced by thefts of non-LoJack equipped cars in neighboring states."[41]

Technically, the LoJack system applies commercially available circuits found in such reference works as John Marcus's *Sourcebook of Electronic Circuits* but miniaturizes them with state-of-the-art solid state technology. Such was the case for both the transponder and the location indicator. From a close read of his patent, it appears that Reagan drew heavily from his previous experience dealing with military and particularly aviation technology. Similar locational technology had been previously worked on by a host of other inventors. For example, in 1967, William M. Silfer Jr. patented a "Method and Apparatus for Locating the Position of a Vehicle" that featured "a mobile radio receiver and transmitter mechanism mounted on an automobile, a station radio transmitter and a receiving and indicating mechanism which are located at a central location, and a pair of radio direction finder stations which are mounted in opposed spaced fixed positions in the perimeter of a selected area." Two years later, Herbert Huebscher patented a "Position Monitoring System" that displayed "motor vehicle positions under multiple path transmission conditions as may exist in a city. At a central station, a signal generator repetitively generates coded interrogation signals for interrogating mobile transponders. Each transponder has timing and transmitting circuitry that transmits a reply signal within a time interval, which is unique with respect to all other transponders in the system."[42]

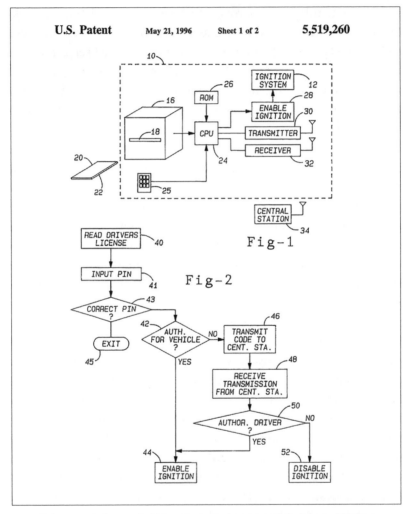

One of the many advanced electronic devices proposed during the 1990s to thwart auto theft. Patented by Valdemar Washington of Flint, Michigan, in 1996, this antitheft system featured the scanning of a driver's license and radio communication to a central data server prior to enabling the automobile's ignition system. The system allowed for multiple users and blocked unauthorized use of a vehicle during specific times of day and night.

While the LoJack, an expensive but effective theft-deterrent option, is not yet available in all areas of the United States, RFID keyless entry systems and vehicle immobilizers are found in most new vehicles manufactured after 2000. Mario W. Cardullo, William L. Parks III, and Charles Walton are credited with the first patents that harnessed these radio wave properties for applications that have ultimately resulted in the RFID automobile key and the labeling of mass merchandise at retailers like Wal-Mart and Target.[43]

Direct applications related to auto security systems, however, awaited the end of the Cold War, the opening of borders in eastern Europe, and a period of massive auto theft during which some 141,000 cars were stolen in Germany in 1992 and more than 1.5 million in Europe. In 1993 the Allianz Insurance Group impressed upon German

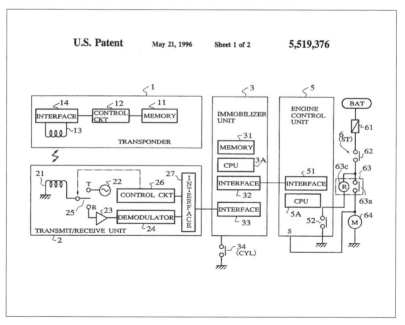

One of the key patents in the development of the RFID ignition key antitheft system. A transponder carrying a unique identification number is embedded in an ignition key. When the key is placed in the ignition, it contacts a receiver that validates the ID number and transmits a signal to an electronic control unit. That unit is connected to an immobilizer box, activating or deactivating the ignition system. Initially the RFID key was thought to be a theft-proof measure, but like other deterrents, it has proved to be vulnerable to smart thievery.

manufacturers the need to improve their antitheft technologies, and that led to the field testing of transponder keys in new vehicles made in Germany and the United Kingdom in 1994. Initial results were nothing less than astonishing.[44] Shortly thereafter, the technology migrated from Ford Europe to Ford's Dearborn operations, where researchers began patenting equipment that subsequently found its way into Ford's PATS (Passive Anti-Theft System), PATS II, and E-PATS systems.[45] The 1997 Ford Mustang had one of the first RFID systems found on American-made vehicles, and theft levels for that model dropped 70 percent when compared to 1995 levels.

At first, manufacturers had every reason to place unlimited faith in RFID security technologies. One early improvement was the expansion of impulses from 32- to 40-bit, thereby increasing code possibilities to the billions. And "rolling" codes were introduced. So at the millennium it appeared that only a thief with a mind like Albert Einstein's would have even a chance to defeat this system and the other new antitheft technologies. But those prognostications were selling human ambitions and cleverness rather short.

Mexico, the United States, and International Auto Theft

5

Borderlines are by their very nature what security analysts call "crimogenic" conditions. Because they divide markets and restrict the exchange of goods and people, creating differentials and asymmetries in cost and incentive and profit, criminal enterprises usually take advantage of these conditions and exploit the asymmetries: auto theft, money laundering, trafficking, black markets, smuggling, prostitution.

JOSH KUH, 2013

Borders—especially national borders—have always been the car thief's best friend. Stolen cars have been transported illegally across international boundaries from the early days of Brass-Era cars to the present. The motor vehicle is a valuable commodity to an enterprising criminal or organization with access to markets that are in disarray and where oversight is hard to maintain. Hotspots abound in the world today, particularly in the war-torn countries of Africa, Central America, and the Middle East, and where there are entrenched black markets such as in eastern Europe and Russia. What is required is the ability to steal, transport, alter the papers for, and dispose of cars across vast distances. Over the years, efforts to thwart international thieves have become increasingly sophisticated, both technologically and organizationally, but so have the methods used by these criminals. It is hard to say who has the upper hand currently, since actors, regions, and techniques are constantly in flux.

This chapter focuses on relations between the United States and Mexico. The long history of auto theft involving these two nations provides insight into how a complex negotiated process has evolved. In addition, the inherent problems facing countries with vastly different cultures, resources, institutions, and markets illustrates the complex picture worldwide. Although the timeline follows historical developments in Mexico, the lessons have far-reaching implications that are relevant to the modern-day context.

A TROUBLED HISTORY OF CARS AND CRIME: THE MEXICAN VICE INDUSTRY TAKES ROOT

The movement of stolen cars into Mexico reflects contrasts between the world's most advanced country and one that has been transitioning for over a century from the corrupting influence of a centralized government to one that is more democratic and decentralized. From the outset, inequality contributed to a well-established vice industry in Mexico that was obvious to American tourists long before cars were commonly seen on its streets. This early vice industry and the subsequent growth of powerful drug cartels provide a window into the trade of stolen cars, since car theft followed the path of organized crime and corruption in Mexico. Within the United States, international auto theft changed over time to reflect the importance of gangs, the significance of technological advancements, demographics, and the rise of regional responses. Prior to the period entering the 2000s, car theft across the Mexican border can be divided into three general periods: (1) before the 1920s, (2) the 1920s to the 1950s, and (3) the 1960s to the 1990s.[1]

THE PRE-1920s: SETTING THE STAGE

Before automobiles defined the landscape, Mexico's vice industry and the U.S. involvement in it were taking root. American entrepreneurs actively invested in activities that were illegal in the United States but located across the border in burgeoning Mexican towns that were accessible by train or horse. Mexico flourished as a haven for gambling, prostitution, greyhound racing, and bullfighting during the Progressive Era (1890s–1920). These pursuits became entrenched in the course of the decade-long Mexican Revolution (1910–20), when the border area was in chaos.[2]

Colonel Estéban Cantú's reign over the northern part of Baja California provides an example of how graft became firmly established in the region. In 1915, in the midst of the revolution, when regional authority often surpassed that of the national government, Cantú issued a gaming permit for the Feria Típica, or Tijuana Fair, that showcased entertainment outlawed in California, including cockfights, boxing, gambling, races, bullfights, and prizefighting. By the end of the year, American promoters were also investing in racing, gambling, casinos, cantinas, bullfighting, and brothels. Cantú benefited both from monthly licensing fees and from smuggling opium, opiates, cocaine, and morphine into the United States.[3] But ordinary Mexicans did not gain from these relationships. Border barons enjoyed a monopoly that excluded Mexican businessmen. Foreign-owned businesses charged less than the prevailing prices, putting local businesses at a disadvantage, and only the lowest-paying jobs were given to local people. One can only surmise that these lawless years included the movement of stolen cars across the border.

THE 1920s TO THE 1950s: FAILED EFFORTS

By the end of the Mexican Revolution (in the 1920s), Americans lamented that Mexico was in disarray, yet the United States had its own social and economic problems. Throughout the Prohibition Era (1920–33), Mexico's vice industry waxed, often financed by U.S. gangsters and racketeers. Conduits for the international transportation of stolen cars were forged. Sinaloa became a hotbed for drug smugglers of marijuana and opium, and it was there that the Mexican cartels got their start.

Nonetheless, attempts to establish formal agreements between U.S. and Mexican officials to recover stolen cars began as early as 1920, initiated largely by the American insurance industry. B. W. McCay, chief investigator for the theft bureau of the Pacific Coast Underwriter's Conference; C. E. DeWitt, chief investigator for the Auto Theft Bureau of Dallas, Texas; J. R. Montgomery, former chief of police of El Paso, Texas; and Major Arturo Kruz of the Mexican government held a conference with national, state, and city government officials of Chihuahua to create a registry of cars owned on both sides of the border.[4] The purpose was to enhance vigilance over stolen cars crossing the border. McCay also reached out to what was then called Lower California with the apparent intent of forging alliances at the regional level across the states of northern Mexico.

While Mexico entered into the postrevolutionary era of nation building, regional cooperation to prevent auto theft continued throughout the 1920s. In a 1925 meeting in Mexicali that included R. H. Colvin of the U.S. Department of Justice; J. E. Ervin, assistant captain of detectives of the Los Angeles Police Department; Walter E. Wood, head of the Automobile Club of Southern California; and General Abelardo L. Rodriguez, governor of Lower California, mutual support was pledged by the Lower California secretary of state, A. Murua Martinez, and Francisco Peralta, inspector general of police from Mexico.[5] Not only were representatives from California invited into Mexico to investigate the ownership of suspected stolen cars, but auto thieves were also deported from Mexico to the United States. In this period of unprecedented joint collaboration, the theft bureau of the Automobile Club of Southern California would telegraph daily bulletins to seventeen hundred police departments, sheriff's offices, garages, and other locations, describing cars that were missing.

The result of these interventions was impressive. The following year, 1926, the *Los Angeles Times* reported that the running of autos over the "line" had been dealt a severe blow. However, the support was hardly one-way. According to the same article, the United States also handed over known "revolutionaries" who were about to invade Mexico in return:

> In expressing appreciation for the close co-operation of the Mexican authorities during the last year which has resulted in very few stolen cars being successfully smuggled across the border, the American representatives were in turn thanked by Gov. Rodriguez for the recent capturing of the "revolutionary army" of seventy to

eighty men about three miles this side of the border shortly before the heavily armed force was to invade Mexico. The trial of the "revolutionaries" apprehended will be held here, with the evidence against them including airplanes, armored cars, machine guns, bombs, and a large amount of other explosives and equipment. A handful of about sixteen United States officials strategically rounded up the entire rebel "army."[6]

Such cooperation continued throughout the 1920s until the Great Depression, when U.S. relations with Mexico soured.[7] During the revolution, about 10 percent of the Mexican population fled to the United States. But after the stock market crash, and for almost a decade thereafter, approximately 458,000 people repatriated back to Mexico, often due to nativist initiatives.[8] Lázaro Cárdenas, who assumed the presidency in 1934, targeted the vice industry and the U.S. "businessmen" who were taking their profits from casinos and other "investments" out of Mexico. Consequently, casinos suffered a serious setback, but corruption had already become firmly established within Mexico.

Throughout the early 1930s, stolen cars taken into Mexico had become a concern of the newly created National Automobile Theft Bureau (NATB), an insurance industry organization. With the help of the U.S. Chamber of Commerce, the State Department, the Senate Foreign Relations Committee, and General Motors, Fred J. Sauter, president of the NATB, arranged a meeting in Mexico City between himself, Ambassador Josephus Daniels, and the Mexican ministers of finance and customs. His objective was to encourage the Mexican government to ratify a pending treaty to return stolen vehicles. His argument to the Mexican officials was that their country was losing the duty on cars stolen in the United States and "smuggled" into Mexico. Sauter later reported: "Just how much this visit contributed to final ratification by the Mexican Senate we do not know, but it is a fact that in less than ninety days after this visit the treaty was formally ratified and put into effect."[9]

In 1936 the United States and Mexico concluded a convention to facilitate the "recovery and restoration of stolen motor vehicles, trailers, airplanes, or the parts thereof," known as the Treaty of Hidalgo.[10] Yet enforcement was fraught with "failure of administrative authorities to conform to express provisions of the Convention."[11] Criminal elements were rampant among the enforcers, as illustrated by the following anonymous communication to the special agent in charge in San Antonio, Texas, dated April 20, 1937; the letter was forwarded to FBI director J. Edgar Hoover.

Sir:

You would be interested to know that the mayority [sic] of the stolen cars in the United States are being brought into Mexico by a bunch of clever crooks, mostly every one of them comes to Mexico via Laredo, N.L. Mexico, and are being helped out by one of the Chiefs of Immigration of that place, his name cannot tell you but it would be easy for you to find out, because this man was here in Mexico City only last week with a fellow by the name of Frank Russell, as I understand they came to find how business were getting along, accomplises [sic] of these fellows are Agustin Dienner who has a Store situated at No. 6 Independencia Street, Francisco Pena an Agente of the Wells Fargo here, this man helps out not only this racket but to bring contraband from Mexico into the U.S. and vice versa of German made guns, and several other things. This man and other fellow by the name of Chapa who has a Store in Santa Veracruz Street help a fellow Clarke E. Clarke, alias Michael C. Clark, Michael C Obrien topass [sic] the cars across the border counting with the help of several politicians and police force there.

Next Door to the Store of the name Chapa is a printing place where all the engraving and letterheads for the bills of lading are done, the name of the owner of this place is Rodriguez and is a friend of the two men mentioned above.

Morton Campbell of Monterrey, who owns a soda Water Factory knows plenty about it, and so does his brother in Sabinas Hidalgo. In Kansas City Mo. Lives a chap by the name of Pete, whose address is 3014 Harrison Street who helps to pick up the cars. Another man is Mike, who has an apartment on 62 Amazonas Street here in Mexico is another of them. Mostly all of the cars when in Mexico City are taken to a Detroit Garage situated in Amazonas Street also.

> Dienner Fellow is a very clever individual and so is this fellow
> Pena. Obrien carries with him a german made gun Parabellum 765
> and so is this other fellow Mike, who is leaving for Texas this week
> and so is Obrien.
>
> Hope this information will be for the good of your country and
> mine.[12]

Stolen cars continued to cross the porous border for decades as the methods for transit and the technology for stealing cars developed into an art. Mexico was also emerging as a low-level but important supplier of marijuana and opium, which became intertwined with car theft. In 1953 an international criminal ring was broken up in which cars were traded for narcotics. The operation unfolded in this way:

> A kingpin heroin peddler had lined up a group of addicts who
> would steal new automobiles on order, submit them to the boss'
> approval and then drive them to Mexicali. The cars were delivered
> either to a gasoline station or ranch on the Mexicali-Tecate road.
>
> The thieves then would return to Los Angeles and be paid off in
> either heroin or cash by El Hombre Uno (The No.1 Man).
>
> If the "hype" (gangland term for addict) merely stole a car he
> would get two grams of heroin If he drove it across the border he
> was paid off at the rate of one-half ounce of the drug or $200 cash.
>
> The receiver in Mexico . . . would deliver the heroin to . . . Los
> Angeles at the rate of from one and a half to two ounces of heroin
> per car.[13]

Another important development related to the movement of stolen cars to Mexico was the post–World War II emergence of gangs in the United States linked to auto theft. Writing for *Police* magazine, retired sergeant Richard Valdemar of the Los Angeles County Sheriff's Department describes how the seemingly harmless indiscretions of the young southern California joyriders became the seedbed of an underworld dealing in stolen cars and car parts: "After World War II automobiles became more plentiful in America and an attitude of tolerance for youthful GTA [grand theft auto] suspects became common. The hot rod and the low rider were often built from parts stripped from stolen cars. These highly modified vehicles were often neighborhood projects built by groups of several young men and teens. Sometimes these cars were built by gangs. About this time also, gangs began to specialize in Grand Theft Auto."[14]

While for the middle-class and affluent youth, the hot-rod era was linked to drive-in movies, drive-in diners, and drag races, for those with less discretionary income, souping up their cars revitalized a home industry for stolen parts. Sometimes it was just groups of people made up of friends, family, or neighbors that hooked up, but other times gangs—or individuals with recognizable names and symbols engaging in illegal activities—were involved. In poor areas, the talents and ties of whole communities were fertile ground for engaging in lucrative ventures in stolen cars. Youths were groomed into car theft at an early age, knowing that if caught they would get a light sentence. Sergeant Valdemar provides insight into how gangs engaged in auto theft:

> GTA is very profitable. Gang run chop shops double and quadruple their profits derived from stolen rides by selling the vehicles' parts separately. In addition, thousands of stolen vehicles worth millions of dollars are smuggled into Mexico each year.
>
> Gang members also steal vehicles to use in other crimes. Burglaries, robberies, and drive-by shootings are often preceded by gang members stealing a car. Car customizing shops are also used by gang members to install hidden stash compartments.
>
> Despite all this GTA activity by gangs, felony GTA charges are rarely filed against these gang members. Unless the vehicle theft can be tied to a GTA ring or a chop shop, the gang member will usually be charged with the lesser crime of "joy riding" because the "intent to permanently deprive" the victim of his or her vehicle is difficult to prove.
>
> Gang members, especially juveniles, receive little incarceration and lots of probation for car crimes.
>
> I think that after the traditional "jump in," carjacking for GTA is the most common gang initiation.[15]

Among the insights Valdemar provides is the historical link between stolen parts and the early years of hot-rodding. The many histories of the hobby are devoid of the importance of this link, which goes a long way toward understanding why the hot-rodding community had such a hard time gaining legitimacy during the 1950s.[16] The issues of that day were about more than just street racing. The rise of gangs and their easy access to Mexico made them ready partners when Mexican organized crime intensified in later years.

The literature on gang formation is large and beyond the scope of this research, but there is reason to believe that youths who join gangs are more likely to engage in car theft than those who don't. For example, when comparing youth gang and nongang criminal behavior, one study found that 44.7 percent of youth gang members engaged in auto theft as opposed to only 4.1 percent of at-risk youths not in a gang.[17] Clearly, not all car thieves belong to gangs, nor do all gang members steal cars, but as antitheft technology made it harder for the casual joyrider to take a car for the fun of it, auto theft became increasingly concentrated in the hands of more hardened elements, as reflected in arrest data that document a decline in youth arrests and an increase in adult arrests.

THE 1960s TO THE 1990s: THE CARTELS AND CAR THEFT

Throughout the 1960s and 1970s, other forces were at play that framed the larger picture. The emergence of the counterculture movement in the United States resulted in a relaxed attitude toward the use of drugs, which in turn broadened the market for illicit substances. Cocaine soon became a drug of choice, fueling the drug trade from Colombia, which was later transferred to Mexico. Luis Astorga and David Shirk explained that "greater U.S. consumption of cocaine in the 1970s and 1980s led to the rise of powerful Colombian DTOs [drug-trafficking organizations], which moved the Andean-produced drug into Miami via the Gulf of Mexico and the Caribbean. As U.S. interdiction efforts in the Gulf gained ground, the Colombians increasingly relied on Mexican smuggling networks to access the United States."[18]

To counteract the rise of drug activity, Mexico created a powerful domestic intelligence agency, the Federal Security Directorate (DSF), which functioned from 1974 to 1985 and was known as the Mexican FBI. The DSF essentially operated above the law, and such immunity inevitably made it a breeding ground for corruption and collusion with the Mexican DTOs that was beyond the reproach of the Office of the Federal Attorney General. A sordid case unfolded from 1979 to 1982 that also revealed a DSF connection with car theft in the United States.

The California Highway Patrol (CHP) first reported trouble in 1979, saying that it was getting a "cold shoulder" from the director of the Mexican Department of Vehicle Registration with regard to a program designed to reduce the number of stolen cars going into Mexico. CHP investigators had discovered both stolen cars and as many as two thousand stolen commercial trucks, vans, and tractor-trailer rigs parked on the streets in Mexican towns.[19] In the previous year, two CHP investigators had visited Cholula, Puebla, Mexico City, and Chetumal and had randomly written down identification numbers, only to find that one in four turned out to be stolen. The lack of interest by Mexican authorities led the CHP to quietly continue its investigation with help from the Baja state judicial police.

In 1981 a *Los Angeles Times* article titled "Car Theft 'Whitewash' on Mexican Side Feared: U.S. Probe of Massive Scheme Points to Involvement by High-Level Officials" reported that since 1975 the FBI had arrested eleven people, including two DSF agents, for the theft of some four thousand high-priced cars from new-car dealerships in southern California.[20] The mastermind was alleged to be Ricardo Rodriguez, one of the two arrested DSF agents, while the primary thief was identified as Gilberto Peraza Mayen. The latter would locate the keys and drive the cars off the dealership lot. He would then turn over the cars to a second driver, who would take them across the border. Also implicated were Mexican customs agents who were allegedly paid off, and a Mexican auto registry official. Other individuals included a former DSF agent who stole cars and delivered them to the head of the DSF in Tijuana. Among the fourteen to seventeen suspects sought by the FBI were the heads of the DSF in Mexico City and in Mazatlán, and Javier Garcia, the son of the national president of Mexico's Institutional

Revolutionary Party (the country's reigning political party for more than fifty years). Garcia also once had headed the DSF in Mexico and served as an undersecretary in the Mexican State Department and as secretary of agrarian reform. It was learned that DSF agents sold the stolen vehicles to high-ranking government officials. During the previous six years, an estimated $30 million to $40 million worth of stolen vehicles had been funneled into Mexico through the crime ring.

Ultimately, fifteen people pled guilty to the charges.[21] During court proceedings, it came out that Peraza-Mayen was taking twenty stolen cars a week across the border.[22] The cars were stolen from dealerships across California—from Burlingame near San Francisco to National City near San Diego—and the vehicles included Chryslers, Porsches, Mercedes Benzes, Volkswagens, and Dodge vans.

The national director of the DSF, Miguel Nazar Haro, was allegedly a co-conspirator. He proved to be a major problem for several reasons. First, according to a 1980 extradition treaty, Mexico was not required to turn over its citizens, so he and the other suspects still in Mexico would likely never face charges. Second, in the course of trying to get Nazar Haro indicted, William H. Kennedy, a U.S. attorney in San Diego, revealed that while Nazar Haro was head of the DSF, he was a source for the CIA.[23] Nazar Haro had resigned his position in January 1982 (to become a business and security consultant to companies in Mexico City), and at that time the CIA supposedly had lost interest in him. But the revelation of his ties to the CIA became a nightmare for Kennedy. Amid calls for his resignation, President Reagan ultimately fired him. In the meantime, Nazar Haro threatened to sue *Time* magazine for libel, since he had been labeled a "thief and conspirator." Yet there was truth to the claim; the *San Diego Union* concluded that Nazar Haro was "at best a thief and at worst . . . a murderer."[24] Nazar Haro testified before a grand Jury in April 1982, thinking he would clear his name, but instead the federal court in San Diego found him guilty of conspiring to transport and receive stolen vehicles.[25] He fled to Mexico, where he was treated like a hero, and never returned to the United States.[26] To Americans, Kennedy was a hero, and in July 1983 he was appointed to the San Diego County Superior Court.[27]

Since the 1936 Treaty of Hidalgo was ineffective in facilitating the return of stolen cars that had crossed the border, and in the midst of the DSF–Nazar Haro car-theft ring affair, an informal system of exchange was established whereby the California Highway Patrol and the insurance industry bypassed the treaty. The result was the recovery of 667 stolen cars in 1980 and more than 700 in 1981. In 1982 Congress held hearings on the subject of improving the exchange of information on car theft between the United States and Mexico, and a revised treaty followed in 1983. However, during the next two decades, a virtual flood of stolen American vehicles crossed the border. One investigator, writing in 1987, estimated that some twenty thousand stolen trucks, cars, and off-road machinery moved into Mexico annually.[28]

Graft was clearly an issue.[29] However, another was Mexico's economic breakdown. Between 1977 and 1981, the country's deficit surged from 20 percent to 57.7 percent of GDP. In 1982, the peso collapsed. Because Mexicans had little disposable income, the market for cars plunged by 41 percent.[30] They flocked to the northern border region in droves, crossed over in search of jobs, and found ways to address the demand for cheap cars back home through auto theft. Border cities became jumping-off points for the distribution of stolen cars. Reports of stolen cars crossing into Mexico from Texas were rather astounding. One observer wrote, "Estimates of southbound thefts presently range from 90 percent of total thefts in Brownsville to 80 percent in El Paso."[31]

At the time, analysts described the workings of professional binational vehicle-theft rings based in such cities as Juárez, Nuevo Laredo, Reynosa, and Matamoros with the caveat that information regarding these rings was limited and came from informants or captured offenders. Part of the problem was direct involvement of the Mexican police and customs officials, because it "both encourage[d] the common employment of stolen vehicles and preserve[d] the public anonymity of ringleaders, allowing them to direct cross-border operations with virtual immunity from legal interference."[32] With the economy in disarray, Mexican police had developed a system of *madrinas* (bridesmaids), a term for nonsalaried underlings, who received their money from *mordidas,* or bribes. *Madrinas* were initially supposed to function as informants and connections between organized crime and authorities.[33] Michael Miller explained: "Auto-theft investigators allege that when Mexican officers are engaged in ring activities, they rarely perform the act of theft themselves, but leave the work to *madrinas . . .* or others willing to risk the arrest."[34] Richard Valdemar described the inner workings of the *madrina* system and its far-reaching effects this way: "About 25 policemen might be regular cops but another 25 are *madrinas* who are not on the payroll. They get their pay by extorting, kidnapping, and robbing and become wealthier than regular cops. If they get caught, Mexico says they're not on the payroll."[35]

It was not uncommon to see Mexican police driving American vehicles that likely had been stolen, tangible presumptive evidence of corruption not only on the part of officials but pervading the entire process of governance.

The rise of car thefts was increasingly tied to the emerging drug trade, as thieves used cars as bribes and for moving products. The dynamic U.S.-Mexican border reflected multiple gang strategies and connections and in some cases sophisticated planning and execution. Michael V. Miller describes it accordingly:

Frontera [border] organizations obtain vehicles from population centers throughout the Southwest. Away from the border, San Antonio and Houston are significant harvest sites in Texas, although *frontera* rings rarely engage in initial theft there. Rather, vehicles in these communities are usually gathered by professional operations, then driven south to the border by aliens returning to Mexico, where they may either be chopped or, more commonly, taken across the boundary by *frontera* rings. A more complex strategy combines southbound theft with northbound theft and smuggling. For example, a minority of cars and trucks stolen in Laredo and the Lower Rio Grande Valley (namely, McAllen and Brownsville) are used to transport drugs and undocumented workers to San Antonio and Houston. Once at destination, other units are then procured from local thieves and driven back to the border. Since several "runners" (return drivers) are often taken on a given northbound trip, four or five vehicles from the Texas interior may go into Mexico for every unit stolen at the border for smuggling purposes. This strategy entails certain risks, but the dovetailing of crime in both directions maximizes profits for *frontera* rings.[36]

In the United States, the police in larger, nonborder cities rarely got involved in the capture of auto thieves caught along the border, even if the car came from their jurisdiction. One police chief commented: "San Antonio and Houston have so much crime, they're not interested in prosecution here. They just notify the owner their car is here and that's it."[37]

For thieves, the crime was one of "harvesting" from a shopping list of cars that became their target. The thieves would strike downtown streets and shopping malls during business hours and flee into Mexico in a matter of minutes. The outbound border crossings were generally unmonitored by U.S. authorities, while the "crush of commuter traffic or the payment of *mordida* [bribes] to *auduna* [Mexican customs agents]" made it easy to enter Mexico.[38]

Within Mexico, the drug trade was also intensifying, and it affected auto theft. In the late 1970s, the Sinaloa drug lords expanded their operations to include Chihuahua and the Texas border. Fellow drug lords relocated their operations to Guadalajara and developed close associations with Honduran cocaine suppliers and the Colombian Medellin cartel. As the Colombian drug-trafficking organizations began unraveling in the late 1980s and early 1990s, Mexican organizations filled the void. The intensity of drug smuggling from Mexico to the United States caused a major backlash by Mexican federal police in concert with the U.S. Drug Enforcement Administration (DEA). War was essentially declared when the Guadalajara cartel took its revenge through the 1985 brutal murder of undercover DEA special agent Enrique "Kiki" Camarena. The founder of the Guadalajara cartel was arrested in 1989, but he continued to run his extensive drug trafficking organization through his former lieutenants. His nephews established the Arellano-Felix organization in Tijuana. The situation was inherently unstable. As one commentator remarked, "A rivalry soon developed between the Sinaloa and Tijuana cartels. Competition for the lucrative plaza, drug gates and smuggler routes soon led to war. This was further complicated by temporary alliances and betrayals with other cartels such as the Gulf and Juarez cartels."[39]

Unorganized alliances between U.S. street and prison gangs and the Mexican drug traffickers, active since the 1920s, were transferred to the Mexican cartels during the 1980s and intensified in the 1990s. Sergeant Valdemar traced the relationship between U.S. gangs and Mexican cartels to the 1990s:

> The Mexican Mafia (members can be any race although mostly Latino) was born around 1990 from California gangsters who made connections with Mexican drug cartels. Cartels operate in the U.S. They sell and distribute drugs through gangs. The gangs are called *sudeños* (southerners) and *norteños* (northerners) in California. Arellano/Felix allied with some, others with Ciudad Juarez or Sinaloa (Chapo Guzman). The Mexican Mafia is doing to the US what the Mexican cartels did to Colombia. They control the distribution and entry points. They are the "gatekeepers" or "plaza" along the routes. Everything must go through them, and they become bigger.[40]

The vast resources of the Mexican cartels made such crimes as car theft that moved vehicles from the United States into Mexico and back again extremely lucrative and difficult to combat. However, Mexico did not have a monopoly on organized crime and auto theft in the United States. Asian gangs, Russian and eastern European mobs, Latin American gangs, and warlords from around the world were all stealing cars. According to a 2005 assessment of national gang activities, approximately 26 percent of law enforcement agencies reported that gangs were associating with organized crime groups in their jurisdictions. Car theft played only one part in this relationship: "These [organized crime] groups often turn[ed] to gangs to conduct low-level criminal activity, protect territories, and facilitate drug-trafficking activities."[41]

Within major metropolitan areas such as Los Angeles, small cities and communities with large immigrant populations known on the street as go-to places for forged green cards also feed into the documents market for stolen cars. Sergeant Richard Valdemar describes it this way based on his experience in Los Angeles County:

> A new problem has arisen in small cities with a large Hispanic and large illegal population like Bell, Bell Gardens, and Hawaiian

Gardens. They've become centers for false documents. They make false driver licenses, social security cards, and immigration documents (green cards) for human trafficking. Their insurance papers and car registrations are super good. In part, it's due to a corrupted DMV. When California released a new format for the California driver's license, counterfeiters were producing false ones before the new ones hit the street. Gangs such as Florence 13, 18th Street, and MS13 are into this.[42]

Also not captured in published statistics is the role of Native American reservations. San Diego County is home to the largest number of reservations in the United States. Reportedly, auto thieves will try to go through reservations when traversing between the United States and Mexico, because the perception is that they are less likely to get caught.[43]

GANGS, YOUTHS, AND WOMEN

As the significance of gangs in international auto theft increased, so did the importance of youths and women. Sociologist Michael V. Miller, in his 1987 study of auto theft along the Texas-Mexico border, described the use of youths in a large Matamoros organization: "This organization, also reportedly engaged in drug smuggling, is thought to include more than 50 members. Matamoros juveniles are recruited, trained, and directed in auto stealing by a cadre of career thieves, who in turn pass vehicles up the gang hierarchy for marketing in Mexico. The ringleader and his top lieutenants thus are well distanced from the act of theft in the U.S."[44]

Youths commonly drive to the location and are the drivers of the stolen cars, while older members are responsible for scouting, the theft itself, and disposal. The young people are typically introduced to the crime through friends and work their way up into more specialized or "prestigious" roles. These gangs did not have to be large to be effective. In a 1996 survey of the San Diego County Regional Auto Theft Task Force staff, 73 percent of respondents estimated that gangs consisted of five or fewer members while the other 27 percent saw the size of gangs as varying or consisting of more than five members.[45]

Whether auto theft is committed primarily by adults or by youths or by a combination of the two engaged in gang activity, it can vary over time and by location. In a discussion of the role of international organized crime and local gangs, Miller describes how location could affect the role of each: "Binational vehicle theft is a crime accomplished primarily by professional rings based in such *frontera* communities as Juárez, Nuevo Laredo, Reynosa, and Matamoros. . . . El Paso stands out by having the lowest ratio of crossborder thefts (as indicated by its relatively high recovery rate) among border cities. This appears to be due to a greater incidence of joyriding, and the presence of many indigenous gangs stealing for the large El Paso parts market."[46]

Women present a different story. When examining specific instances of auto theft across the U.S.-Mexican border, sociologist Rosalva Resendiz identified the role of women as secondary and situated it within a recognizable division of labor:

> In order to have a successful theft, *robacarros* [auto thieves] developed a division of labor. This division of labor is characterized by three roles: the *chauffeur,* the *specialist,* and the *mounter.* The *chauffeur* drives his or her own vehicle and is responsible for driving the *specialist* and the *mounter* to the targeted vehicle, and picking them up after completion of the theft. Once the *specialist* and *mounter* are dropped off, the *specialist* is responsible for "breaking into" and starting the vehicle. The *mounter* then drives the vehicle to Mexico for delivery. . . .
>
> The role division of auto theft can be recoded into two general codes: the *amateur . . .* and the *specialist.* The *chauffeur* and *mounter* role[s] can be compiled into the role of an *amateur,* which is a secondary role requiring low levels of skill. The *specialist* was the professional auto thief with the highest level of skill.
>
> Differences between the *amateurs* and *specialists* emerged in gender, and age distribution. Females were concentrated as *amateurs,* having started their professional auto thief career as adults, while the *specialists* began their career as juveniles. *Specialists,*

who possessed certain skills, tended to see their deviant activities as an "occupation."[47]

Because the female auto thief encounters many male figures in the commission of the act and the disposition of the auto, it may be particularly difficult for her to break through the multiple gender barriers along the chain of command. However, that is clearly changing, as women are now beginning to head cartels and take more prominent positions in the cartel structure.[48]

Seen as a whole, the organization of gangs reveals several roles in the theft of cars that are generally assumed by different people with specific knowledge:

1. Thief—specializes in types of vehicles. If in an organized gang, may have expert knowledge of electronic car protection systems.

2. Forger—buys and sells documents (stolen or blank) and forges and alters papers

3. Ringer—re-VIN a stolen car. Usually car technicians or mechanics.

4. Courier—transports the stolen car

5. Seller—hired to market the vehicles

6. Coordinator—recruits thieves, forgers or ringers or acts of a thief, forger, ringer or seller

7. Organizer—supplies the initial capital and takes a commission.[49]

Gangs of highly skilled thieves are needed for a procedure that has recently become a trend: stolen vehicles are taken into Mexico to be outfitted with phony registrations, counterfeit identifications, and license plates, and then they are returned to the United States for sale.[50] Clearly, the growing sophistication of international auto theft has required that law enforcement develop a unique set of skills and an organizational capacity to combat these well-equipped perpetrators.

UNINTENDED CONSEQUENCES OF THE UNDERGROUND MARKET FOR CARS

For many years Mexico, like other Latin American countries, banned the import of used automobiles to protect its own car market. This closed-market system stifled competition and drove prices up. Consequently, people purchased cars in the United States and drove them into Mexico, relying on what many say was corruption in customs posts and the courts; in Mexico the vehicles were sold without proper documentation.[51] Buyers preferred these cars because they were much less expensive and because the buyers were spared taxes and registration, but the cars were often in very poor condition. A typical car was bought at a junk auction for $500–$600, fixed up, and sold for as much as $1,500. Such underground cars were referred to as *coches chocolate,* or "chocolate cars," a name that allegedly came from a line in the Tom Hanks movie *Forrest Gump*—"Life is like a box of chocolate—you never know what you're gonna get."[52] In any case, the smuggled vehicles resulted in a lively black market.

By 2000 Mexican roads were cluttered by an estimated 3 million illegal foreign automobiles, including many stolen ones—likely hundreds of thousands of them.[53] The situation was ripe for exploitation. By some accounts, owners of such vehicles were forced to pay bribes to police and other authorities to keep driving them.[54] An example of how widespread the problem had become is that Chihuahua farmer and social activist Julian LeBaron estimated that one-third of the vehicles in the state of Chihuahua were "chocolates."[55]

This irregular process continued until 2005, when President Vicente Fox authorized a plan that would legalize foreign used vehicles more than ten years old. The initial reaction was that this was beneficial to "chocolate car" owners, but other problems, including issues of environmental pollution and detrimental effects on the domestic auto industry, have surfaced over time. And while it was not intended to result in the legalization of stolen cars, some thieves likely benefited by the action. Later, in 2008, President Felipe Calderón issued an Automotive Decree that brought the importation of used cars from the North American Free Trade Agreement (NAFTA) region more into compliance with NAFTA provisions.

INSTITUTIONAL AND TECHNOLOGICAL METHODS TO COMBAT AUTO THEFT IN A TRANSNATIONAL ERA

The long history of auto theft between Mexico and the United States includes efforts to keep track of stolen cars, develop cooperative arrangements, and understand the criminal links across borders. Consequently, beginning in the 1990s, the United States focused on means to combat auto theft by adopting aggressive measures to counter the growing violence, developing databases and creating new forms of interagency communication. The 1992 Anti–Car Theft Act made armed auto theft, or "carjacking," a federal crime. Two years later, the Violent Crime Control and Law Enforcement Act made carjacking resulting in death a federal crime punishable by death. But perhaps even more significantly, that act also required businesses and governments to check VIN's against the FBI's stolen-car database and certify that salvaged or junked vehicles were not stolen; furthermore, the law provided for startup funds to link all state motor-vehicle departments and a grant program to create state and local antitheft committees. These provisions laid the groundwork for developing database clearinghouses. Among them were the National Motor Vehicle Title Information System and the National Insurance Crime Bureau.

These national systems also supported the efforts of regional institutions begun in the 1980s, known by such acronyms as ACT, ATPA, HEAT, and CAT. In the 1980s states and regions experiencing high auto thefts began to form Anti–Car Theft (ACT) groups funded by grants from coalitions of law enforcement groups, state funds, insurers, and consumers to promote public awareness of vehicle theft and lobby for passage of state legislation aimed at combating thefts. At least thirteen states (Arizona, California, Colorado, Florida, Illinois, Maryland, Michigan, Minnesota, New York, Pennsylvania, Rhode Island, Texas, and Virginia) had created Automobile Theft Prevention Authorities (ATPAs), mostly funded by a small surcharge on drivers' licenses or

Operation C.A.T. was a regional response to auto theft in California in the 1990s. This decal, when found on a vehicle's rear window, authorized police to stop the car without cause after 11:00 p.m. Photograph by John Heitmann.

registration fees or on auto insurance policies sold in the state. Michigan pioneered the ATPA concept in 1986, allocating one dollar from each auto insurance policy and channeling the funds toward combating auto theft. Michigan's program called Help Eliminate Auto Theft (HEAT) includes a hotline for residents to report thefts and chop-shop operations. In the twenty-five years since the program was instituted in 1985, information forwarded to HEAT has resulted in the recovery of more than 4,200 vehicles, valued at more than $51 million, and has led to the arrest of almost 3,400 suspects.

ATPA's and other state entities such as ACT groups use a wide range of programs to fight auto theft. Besides HEAT hotline programs, Combat Auto Theft (CAT) programs involve auto owners who voluntarily put stickers on their windshields that alert police that they can stop the car for a theft after a certain hour. High-theft metropolitan areas have instituted task forces to combat auto theft. In Newark, New Jersey, a task force helped reduce the city's theft rate from the highest in the United States in 1991 to sixteenth in 1996.[56]

One of the Automobile Theft Prevention Authorities is the San Diego County Regional Auto Theft Task Force (RATT). This RATT was formed in July 1992 after years of escalating thefts, violence, and Mexican ties during the 1980s. A 2000 internal document noted that "from 1984 through 1989, motor vehicle theft rose 151 percent" in the San Diego area.[57] By 1996, San Diego was second only to Los Angeles in the number of cars stolen in California. The increase was attributed to ties to Mexico.[58]

As of 2000, the California Department of Motor Vehicles paid the salaries of twenty-one RATT investigators from several police agencies, three deputy district attorneys, and three support staff. The funds are administered by the county district attorney's office. Five FBI agents and one agent from the National Insurance Crime Bureau (NICB) brought the total number of investigators to twenty-seven.

As in other regional programs to combat auto theft, RATT includes entities that are not considered to be traditional law enforcement agencies. In this case, sixteen federal, state, and local law enforcement agencies participate in RATT: the FBI, the U.S. attorney, U.S. Customs, the California Highway Patrol, all ten local police agencies, the San Diego County district attorney, and the NICB.

RATT combines "traditional enforcement strategies (e.g., surveillance, search warrants, arrest warrants, sting operations using covert warehouses as 'chop shops,' and videotaped buy-busts" with investigative techniques:

1. Theft analysis—by tracking locations and types of vehicles stolen and monitoring known "chop shops"

2. Maintenance of an intelligence database—by utilizing data from the California Law Enforcement Telecommunications System, the National Crime Information Center, and National Insurance Crime Bureau

3. Active liaison with all local law enforcement agencies

4. Recruitment, development, and careful supervision of informants—to infiltrate car theft rings using informants and undercover tactics.[59]

RATT and other regional agencies illustrate the dependence on "information-led" attempts to combat auto theft. Such a tactic is particularly important in the four states bordering Mexico, which account for one-third of the vehicles stolen in the United States. An additional tool is "bait vehicles" that use audio and video devices to improve communications, tracking, mapping, and remote-control technologies. Together, they represent a growing trend by law enforcement agencies to depend on information technologies as they fight against an often sophisticated and lethal opponent.

Within each state various layers of agencies operate in combination with federal or local governmental units, private-sector organizations, and the public to combat auto theft. The state of Texas provides an example of the complexity. Texas legislation established the Texas Automobile Theft Prevention Authority in 1991 to address the growing problem of auto theft. The authority was expanded in 2007 to include auto burglary, and the agency was renamed the Automobile Burglary and Theft Prevention Authority (ABTPA). The broad mandate of the ABTPA includes law enforcement, detection, and apprehension; border enforcement; public awareness and crime prevention; antitheft devices and registration; and statewide training enforcement and detection. According to a document summarizing its activity, "ABTPA functions as the lead organization in a statewide network of law enforcement agencies, prosecutors, insurance industry representatives, local tax assessor-collectors, community organizations, and concerned citizen groups."[60]

To deal with the cross-border issues, in 1992 ABTPA created the Borders Solutions Committee (BSC) with representatives from the Texas Department of Public Safety, the General Services Administration, the NICB, the El Paso Police Department, the El Paso Sheriff's Office, the Texas Department of Transportation, the Allstate Insurance Company, the Mexico-Texas Bridge Owner's Association, and the Texas Attorney General's Office.[61] Among other things, the BSC fostered cooperation and communication between the two countries. In September 2000 the BSC launched the Border Partners Program, which created a cooperative approach along the El Paso–Juarez border for repatriating stolen vehicles. This regional initiative "assists auto theft victims and manages data in the computer network which links several bordering states to recover vehicles from Mexico. Authorities on both sides of the border access the same database of stolen vehicles and jointly investigate and prosecute auto thieves."[62]

Yet another initiative of ABTPA is the Border Auto Theft Information Center (BATIC), created in 1994 to promote partnerships and maintain data. According to ABTPA, prior to BATIC, U.S. law enforcement recovered fewer than 500 stolen vehicles annually from Mexico. Since BATIC's inception, however, the average recovery rate has mushroomed more than fivefold to 2,600.[63] In conformance with the Hague Convention on the Taking of Evidence Abroad, which provides for cross-national cooperation in training, workshops, seminars, and field exercises, BATIC partners with the NICB in conducting law enforcement training in other countries. Although the partnership was initially focused on relations with Mexico, it has expanded to include Canada as well as certain countries in Central America and the Caribbean, although Mexico is by far the major partner (see table 5.1).

According to BATIC, the states reporting the most stolen vehicles located in Mexico and Central America are (in descending order): California, Texas, Arizona, Colorado, Florida, New Mexico, and Nevada. The states within Mexico where the most vehicles stolen from the United States are located are (in descending order): Baja California, Chihuahua, Sonora, Tamaulipas, Sinaloa, and the Distrito Federal (Mexico City).[64] It is surprising that such an important program operates only in Texas. Arizona attempted to create a similar agency with Sonora but failed owing to funding constraints. BATIC has become crucial to U.S. endeavors to combat cross-border auto theft. Further, it is a dynamic organization. The Texas Department of Public Safety indicates that "the use of B.A.T.I.C. is becoming more and more evident in the United States and remains the only program of its kind. Over the last few years, it has evolved to keep up with new techniques in discovering stolen vehicles to better assist law enforcement officers on both sides of the border."[65]

The ability to engage in successful international programs in the identification and recovery of automobiles depends on having partners that are politically stable and able to enforce the rule of law. Clearly, this has been difficult with the spreading influence of Mexican drug cartels and the cross-border liaisons of violent street gangs such as MS13, which originated in El Salvador, migrated to the United States, and returned home with a vengeance. Efforts by foreign governments to clean up law enforcement and go to war with cartels and gangs have been essential to stolen vehicle recovery in the United States.

Before the regional and international agencies were established in the 1990s and the early years of the 2000s, the methods of vehicle recovery depended upon informal cooperation or formal treaties. In what was perceived as the more successful informal approach, a designated officer acted as a liaison with local Mexican officials to secure the return of stolen vehicles. The liaison officer was a broker who generated goodwill and ensured cooperation with authorities by providing such favors as crossing documents and consumer goods.[66] It was the faster of the two approaches and the one most commonly used, but by essentially operating under the table, it merely perpetuated a corrupt system. It is described in the following way:

> Informal recovery begins when Mexican police inform the liaison officer that a unit is available for return. After identifying the vehicle, the liaison contacts the owner, which by this time is usually the insurance company. The insurer, in turn, subcontracts recovery to one of several local independent adjusters specializing in the task. . . . Cash payment releases the vehicle to the adjuster, whereupon it is towed back across the border to a salvage pool destination and eventually auctioned to used car or parts dealers.
>
> While not universal, *mordida* [bribe] is the norm. All parties viewed it as virtually indispensable to informal recovery. The precise amount extended appears to reflect the estimated value of the unit and the agency in possession.[67]

The alternative, or formal approach, required that local law enforcement agencies recover stolen vehicles through treaties. Without a treaty, the recovery of stolen vehicles was and continues to be difficult. If no treaty was in place, the laws of the country could favor the foreign purchaser rather than the U.S. owner. Even if a stolen vehicle was identified abroad, the release petition required more documentation than is needed under a treaty to meet international legal standards. In addition, the process for recovery could change at any time.[68] From the American perspective, the process was cumbersome and could take several months from the time a stolen vehicle was identified as held by Mexican authorities. From the Mexican point of view, authorities with salaries as low as $180 per month and limited resources had few incentives to cooperate.[69] The situation did not change until the rule of law in Mexico began to change.

Nonetheless, the United States has continued to forge similar bilateral treaties with Central American and Caribbean countries, including the Dominican Republic (2001), Panama (2002), Belize (2002), Honduras (2004), and Guatemala (2007), along the lines of those established with Mexico in 1983. At a minimum, they establish procedures for the recovery and return of stolen vehicles. In addition, the NICB, though not a party to the treaties, assists U.S. consular officers by checking its database and that of the FBI's National Crime Information Center to determine whether vehicles have been reported stolen. The institutional mechanisms for identification and recovery create important scaffolding for moving the process forward.

However, there remains a paucity of treaties between most first-world and third-world countries. Arrests of foreign nationals in the United States indicate how vast the problem has become. A 2011 three-month nationwide sweep of gang members with ties to drug-trafficking organizations who faced charges ranging from drug and parole violations to vehicle theft included people from Russia, Thailand, and Laos as well as from Mexico.[70] Also in 2011, three Iraqi men were arrested in El Cajon near San Diego as part of an international auto-theft ring based in the Detroit area for vehicles driven to Canada and shipped to Iraq.[71] The Iraqi connection is particularly troubling given the possibility that stolen vehicles could be used to both fund terrorist organizations and deliver weapons of mass destruction such as improvised explosive devices.[72] Clearly, the treaty process can't keep up with current events.

In the late 1970s and early 1980s, the U.S. Department of Justice delegated auto-theft crime enforcement to local crime agencies. Since then, specialized authorities have emerged across the country. An example is the California Highway Patrol's Foreign Export and Recovery Unit (FEAR), formed in 1995, which monitors the movement of stolen goods, including vehicles, from the ports of Los Angeles–Long Beach and the San Francisco Bay area. FEAR operates in conjunction with other governmental agencies such as the U.S. Customs and the Federal Maritime Commission, as well as the NICB. The height of export operations was in the 1990s; most current crimes are focused on title-washing and fraud.[73] The Border Division of the California Highway Patrol also runs a satellite office of FEAR that intercepts stolen vehicles headed south for sale in Mexico and Central and South America. In addition, a Mexican liaison unit works with Mexican authorities to recover stolen vehicles that make it across the border.

Other countries have also created their own auto-theft recovery agencies, such as Mexico's Oficina Coordinadora de Riesgos Asegurdos S.C. (OCRA), or Office to Coordinate Insured Risk. An organization representing insurance companies and headquartered in Mexico City, it is focused on the repatriation of stolen Mexican vehicles into the United States. Car insurance is expensive in Mexico, and many drive without it. Consequently, criminals have found it easy to steal cars or engage in carjackings, whether or not the car is insured. This form of auto theft is estimated to have increased dramatically in the past five years. Although cartels might not be running the car-theft operations directly, most car-related crimes are commonly thought to be linked to them.[74]

However, during the recent past, belt-tightening by governmental agencies, some states, and local entities has led to the elimination of vehicle-theft units to cut costs. But these closures come at a high price. Reportedly, it takes about five years for an investigator to learn advanced vehicle identification.[75] In addition, there are the intangibles related to knowing whom to contact in another country. The fear expressed by Christopher T. McDonold, former president of the International Association of Auto Theft Investigators and detective with the Baltimore County, Maryland, Police Department, Regional Auto Theft Team, is that as local agencies expend fewer resources on auto theft, international criminal elements will fill the void. McDonold stated in 2011: "In current and future trends, experts expect auto thieves to become more technological, employ more sophisticated schemes, and transport more vehicles across international lines."[76]

THE JOYRIDE IS OVER

It would be heartening to suggest that car-theft levels are down, as indicated by the data, but in fact the new faces of crime involve fraud, which is usually not reported as a stolen vehicle. A vehicle stolen from a car dealer, cloned, given a new title and other paperwork, and exported, is an instance of fraud and may not show up in the statistics as theft. The definition of car identity theft can vary from one law enforcement agency to another. For example, the Los Angeles Police Department will not take theft reports owing to fraud, but the Los Angeles County Sheriff's office will. Consequently, thefts that involve sophisticated methods for disguising the identity of the vehicle may be masking the full extent of the problem.

In the current era of integrated economies, a seemingly unrelated action such as the devaluation of the dollar can affect the market for stolen U.S. cars abroad.[77] Ken Ellingwood and Tracy Wilkinson explain:

> One of the most common reasons for vehicle theft is the ability to generate profit. . . . The answer for many questions about vehicle theft may be found in financial news sources. According to a recent article in *USA Today,* "A weaker dollar has made exports from the United States more desirable in many countries such as Canada, Mexico, Saudi Arabia, China and Germany. More than 1.5 million new cars were exported last year, up 38 percent from 2009, Commerce Department data show. Last year's automotive exports were valued at $36.7 billion." If there is a market for new vehicles, think of the market for stolen vehicles.[78]

Yet another concern has emerged since the means for combating auto theft associated with cartels has reached a new level of violence. Drug traffickers have reportedly acquired military-grade weapons including grenades, grenade launchers, armor-piercing munitions, and antitank rockets.[79] Thwarting international auto theft now requires being part of a coordinated strategy for combating far-reaching major criminal activities.

Seen as a whole, the methods for stealing cars, the people involved, and the markets are far more complicated than in previous eras, have multiple facets with implications from monetary policy to drug policies, and require new levels of integrated international intervention involving both governmental agencies and private organizations on an unprecedented scale. Anything less makes it harder to keep pace with the problem, to say nothing of meeting it head on.

Not only are the perpetrators more skilled and more integrally linked into international networks than in the past, but the underground society also feeds off a culture of permissiveness. *Narcocorridos,* or narcotics trade ballads, and *narconovelas,* or narcotics trade novels that glorify drug traffickers, are pervasive in Mexico and parts of the United States. A recent *narcocorrido* posted on YouTube essentially taught viewers how to steal a car[80]—one more indication of how cultural representations of auto theft can affect individuals' actions.[81]

6

The Recent Past

I have discovered you have to work twice as hard when it's honest.
SARA "SWAY" WAYLAND (ANGELINA JOLIE), *GONE IN SIXTY SECONDS* (2000)

In the closing scenes of Clint Eastwood's 2008 film *Gran Torino,* the stories of one man's personal redemption and another's dream of achieving independent manhood come together in two life-defining moments: one of self-sacrifice, and the other a symbolic act of automobility.[1] Confronting a gang that had terrorized his newly adopted family of immigrant Hmong neighbors, the cantankerous Polish-American autoworker and Korean War veteran Walt goads the thugs into murdering him before witnesses and thereby saves the community. By his death, Walt spares the life and ensures the innocence of Thao, the neighbor boy who was intent on exacting revenge for the rape of his sister by the gang. For Walt, the thought of the good he is doing may ease the haunting memory of his killing of an enemy prisoner in Korea. Thao is his last chance at redemption. Thao, whom Walt had guided in the previous months into self-respecting appreciation of hard work, independence of mind, and success with the ladies, is last seen driving Walt's beloved Gran Torino toward what must be presumed to be a future life of dignified manhood. This story of tragic nobility takes place in the "motor city"—Detroit, Michigan (more specifically, Highland Park, where Ford Model T's were first built). And the story begins with an attempted theft by Thao of Walt's Gran Torino.

The car featured in the film—a Ford Gran Torino SportsRoof—was a vestige of the glory once associated with "Detroit Iron." The green muscle car featured body-on-frame construction rather than a cheaper unibody design, along with a long hood that had a scoop and a short deck. Most commentators of the day thought the car looked good, and interestingly, it handled far better than its competition.

What is the essential significance of Walt's Gran Torino? On one level, the car was a catalyst for what followed, and nothing more. But the type of car, a 1972 Gran Torino—American muscle made at the end of the nation's love affair with the car—was the last thing of loving importance in Walt's undistinguished working-class life. Walt had just lost his wife, and his family was totally disaffected toward him. His neighborhood was no more—he was one of the last white Americans living in it, the others all having fled. Walt's car, still looking like new, represented a world long gone, one on which we look back with nostalgia. Thus, the Gran Torino signifies dignified independence, covered in a garage, while Walt uses a rusty old truck as his daily driver. And why would one drive the Ford in a world gone wrong? Walt is waiting for a worthy new owner, certainly not from his own family. It is Thao, whose independent actions result in his becoming a true American, who is now worthy of getting behind the wheel of this American classic. And while this story illustrates the connection between auto theft and the muscular American male, in more recent times the act of auto theft has been increasingly associated with the intelligent and technologically adept male as well.

BETTER TECHNOLOGY, SMARTER THIEVES

In 2010 John R. Quain, writing in the *New York Times,* summarized recent developments in auto-theft technology this way: "Technology is getting better[;] professional car thieves have stepped up their game, too, meaning that some tracking systems may be better than others."[2]

Without question, technology has had an impact on the decline of theft rates experienced since 2005. And while the joyrider is nearly extinct, professionals continue to thrive, as current recovery rates are at the alarmingly low level of less than 50 percent. Since September 11, 2001, the United States has concentrated its border efforts far more on what comes into the country than on what might go out. Consequently, with law enforcement and customs officials stretched thin, ethnic gangs flourish in the hot-car export markets. Coupled with a high level of insurance-fraud activities and vehicle cloning rings, they give authorities—hampered by personnel and funding cutbacks—more than they can handle.

One deterrent has been General Motors's OnStar, installed in a growing number of vehicles. OnStar is mainly known as a motorist response system for emergencies or accidents. But it also has its own vehicle-recovery program, in which the victim of a car theft calls the authorities and OnStar.[3] Control-center personnel at OnStar then send out a radio wave that can disable the car by preventing its ignition from starting, or it can energize the car's lights to flash and its horn to honk. OnStar personnel can remotely make the vehicle come to a gentle stop if it is already in motion, thus avoiding a potentially dangerous police chase. But OnStar has its flaws. For example, during the time lapse before the owner of the boosted vehicle reports the car as missing, thieves may be able to either strip it or simply disable the equipment. The Internet is also full of advice, some probably good, some bad, on how one might defeat the OnStar system. One website instructs a would-be thief to find the unit under the front passenger seat and then unplug terminals J1, J2, and J3, while keeping J4 connected to the unit.[4] A so-called expert posted the following: "See that little black rectangular box at the right center top of the windshield? Find the feed wire to it, [then] use a small wire clipper to 'interrupt' communications. Permanently." For all units that depend on GPS, just putting the car into an enclosed space serves to shield the unit from a satellite transmission. But if one wishes to use a more elegant technological approach, the GPS Jammer may be the tool of choice. The Jammer 08, available from a Chinese company for $150 on the Internet, prevents cars and the people in them from being tracked. Its makers claim: "It adopts the technology of interdiction and interposition code, so it will intercept the signal of satellite and break it completely."[5]

Thieves can counter even some of the most sophisticated new antitheft technology devices.[6] For instance, the SD-98 device serves as a remote-control unit that can be manipulated to defeat keyless door locks. Essentially, it mimics one's keyless door opener. By activating a master remote control to operate a television or a DVD player, the

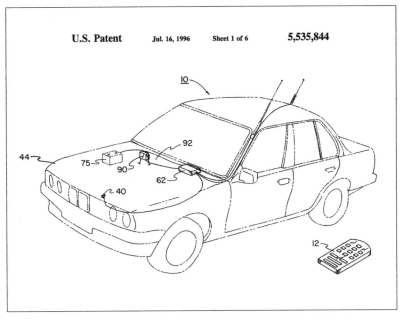

U.S. Patent   Jul. 16, 1996   Sheet 1 of 6   5,535,844

Remote alarm system, 1996. Alarm systems have evolved continuously since the 1930s. Early designs usually featured a device that responded to the bumping or shocking of a vehicle; as they became more sophisticated, hood, trunk, and door switches were connected to a central unit or "brain" that disabled the starter. Ricky Samford patented this remotely activated unit, which featured a radio transmitter and receiver that could cut off the flow of fuel to the engine.

SD-98 auto remote-control blocker works on 868, 433, 315, 305, and 330 MHz. It has four functions—scanning, blocking, jamming, and also operating if the manufacturer's code is known, much like the device it is attempting to replace. It scans for antitheft-device frequencies; when it finds one, it opens a vehicle door without destroying the lock. It also can block a remote-control signal, if there is a key sending such a signal, and then open the car. It can also jam or disable a car key. And finally, if the thief has a manufacturer's code, it opens the door directly. The code works on Audi, Alfa, BMW, Bentley, Mercedes-Benz, Citroën, Fiat, General Motors, Honda, Jaguar, Mazda, Volkswagen, Mitsubishi, Nissan, Peugeot, Renault, Seat, Skoda, and Toyota, up to the model year 2010.[7] And despite the relative effectiveness of the LoJack and the claim that its strong radio signal is difficult to jam, a Chinese company currently markets a "LoJack Jammer."[8] Features of the RMX02 LoJack

4G XM Jammer include disabling the LoJack tracker. The device operates at a range of 5 to 15 meters and is able to produce untraceable RF signals.

WHO CARES?

More recent film and literature lionized the auto thief in a manner that painted the act as largely victimless, harmless to human health, and at times actually comedic. The 2007 independent film *The Go-Getter* paints a picture of West Coast kids living aimlessly and in angst.[9] High school senior Mercer (Lou Taylor Pucci), who recently witnessed the death of his mother, has just read Mark Twain in an AP English class and wants to journey on "the river." Impulsively, he steals a 240 Volvo station wagon at a car wash in his small Oregon town and begins an odyssey to find his half brother, Arlen. Along the way, he discovers that his brother is a no-good and that humans generally are disappointing. There is one exception, however. The owner of the car, Kate (Zooey Deschanel), has left her cell phone in the car, and she begins to have an extended conversation with the young thief. From time to time, she calls Mercer. Ultimately, her empathy with him results in a profound relationship that bonds Mercer with the stolen vehicle's owner. She raises an important question: is "life random or fate?" Was Mercer's theft of the Volvo a totally random act, or was there a deeper meaning behind this crime? Mercer's trip takes him (and for a time a girl living in Fallon, Nevada, whom he once knew in middle school) through such places as Shelter Cove, California; Reno, Nevada; the Mohave Desert; Los Angeles; and finally, Ensenada, Mexico, where he catches up with his half brother and realizes Kate's love. Mercer's next stop is Pointe Coupee Parish, Louisiana, where an aunt and two cousins live, but this time Kate rides along for the drive. In sum, the film reflects a generation's fears and challenges, and the stolen car is a mere conveyance. The old Volvo is nothing more than an appliance, so typical of the way the postmillennial generation views automobiles in general.

In recent fiction also, moral relativity related to auto theft appears to be the "new normal." Janet Evanovich's *Motor Mouth* is an entertaining but forgettable story of sex, cheating on the NASCAR circuit, theft, and the detection of a microchip that is at the heart of a high-tech traction-control system. After a race is lost under suspicious circumstances, the central character in the novel, Barnaby, a woman mechanic who loves pink and was once the lover of driver Hooker, decides that something was not right with the winning car and perpetrates a boost of a hauler with two cars—and, incidentally, a dead body packed in ice. But a GPS system has to be disabled, and Barnaby does that rather simply: "I was able to squeeze my arm far enough to reach a ball of aluminum foil sitting on the kitchenette counter. I ripped a couple of chunks off the roll, swung out of the hauler, and climbed onto the back to the cab. The antenna had been placed in the usual location between the exhaust pipes. I wrapped the antenna in aluminum foil and jumped off. Turns out it's pretty easy to screw up a GPS system." With that clever act, the hauler and the race car are stolen, and the puzzle of why a competitor's car turned out to be so fast is ultimately solved.[10] While the theft is central to the story, it merely enables the author to develop characters with postmillennium moral values.

Finally, Pete Hautman's *How to Steal a Car*, published by Scholastic Press in 2009, was aimed directly at teenagers. The back of the dust cover says it all:

> Are you bored out of your mind?
> Sick of your friends and family?
> Wish you were somewhere (anywhere) else?
> Stealing a car might help.[11]

Written about teenagers for a teenage audience, *How to Steal a Car* is a tale of a fifteen-year-old girl, her friends and schoolmates, and her dysfunctional family. But this is not just any fifteen-year-old girl, for Kell, the novel's protagonist, is a thoroughly middle-class nerd who reads *Moby-Dick* and steals cars. At first it is just impulsive, and then her activities move into the sphere of an organized car-theft ring. As the author interjects, "Most people think car thieves are squinty-eyed young guys with tattoos and grease under their finger-nails, but you never know who will steal a car."[12] Kell lives in a world with few absolute moral standards, little compunction about breaking the law when no one is visibly hurt, and a fiercely independent streak also found in her parents and grandparents. Her conscience rarely bothers her; she relates, "I should say something about my mental state during all of this: Happy and Relaxed."[13] Since she is fifteen, she knows the law will come down easy on her when she's caught, if it does at all, and thus fears little concerning any consequences. As the novel ends, Kell concludes: "I think a lot of car thieves just like to steal cars and drive. Also, they think they will never get caught even though most of them eventually do and they know it but they just don't care."[14]

Contemporary film revealed the youthfulness and societal complexities that were often involved in auto thievery. At times the act was casual and a spur-of-the-moment thing; other times it was part of a calculated plan to steal expensive cars for monetary rewards. For the most desperate, it was either a break from boredom or part of a survival strategy. And with few exceptions, simplistic story lines did little to dig deep into personalities or motives.

CAR THEFT AS A GAME

Electronic gaming, which in terms of profits had far outstripped film by the early twenty-first century, took auto banditry to far darker and more violent levels. Both forms of media brought the viewer or participant into imaginary worlds of entertainment, but electronic gaming was far more intense, emotional, and controversial.

In 1997 car theft entered the digital world in a significant way with the introduction of *Grand Theft Auto* (*GTA*) video games. In the first decade of the twenty-first century, *GTA* was the best-selling and among the most technologically sophisticated games in the competitive video-game industry. "In its ambition, fearlessness, style, and production quality," one reviewer wrote in 2009, "it stands apart from every other franchise."[15] Take-Two Interactive and Rockstar games have sold more than 80 million units of *GTA* and its spin-offs.[16] The action and scope of the digital map, along with driving and gunplay, have given the *GTA* series a strong consumer appeal. Between the introduction of *GTA I* and the release of *GTA III* in 2002, Rockstar transformed the games from a structured set of missions with a top-down birds-eye view of the car into a nonlinear sandbox playground, giving the *GTA* player the freedom to pursue organized crime or, with weapon and automobile, create mayhem.[17] With *GTA III,* subsequently refined with the release of *GTA IV* in 2008, the digital landscape was converted into what the video game world calls a "sandbox"—gamers could, at their discretion, follow *GTA*'s narrative or could drive their stolen automobile around the open digital city. In this digital world, driving is essential to the player's criminal success, and car theft becomes a necessary prelude to other criminal tasks. To complete *GTA*'s narrative, the gamer must accomplish a series of criminal underworld missions. In *San Andreas,* for example, to complete the mission "Life's a Beach," the player must win a dance contest and then steal the "Sound Van" from a local DJ and successfully transport it to a local parking garage.[18] The virtual universe of *GTA*'s "urban action" game revolutionized the video game industry. Importantly, automobile theft and automobile-related violence is, in almost all sequences, the pivotal and most thrilling dimension of the *GTA* experience. With *GTA,* car thieves became one of the most popular avatars in the video game industry. Unlike games with hero-avatars who eliminate bad guys for a self-proclaimed righteous cause, the *GTA* player controls criminal-avatars who carry out illegal tasks or, if the player chooses, commit random violence on innocent bystanders and pedestrians. Players assume the criminal's identity; they see the game's digital world through his eyes. The digital criminals can—at the player's discretion—assume one of the automobile thief's many personas: the youthful joyrider, the professional thief, the carjacker, the reckless escapist, the drive-by shooter, the placid cruiser, or the savvy criminal who, in a stolen car, commits murders, deals drugs, or kidnaps.

Each version of *GTA* has a particular criminal ethos, intimately connected to automobile theft. *Vice City* (2002) is set in a fictional Miami, and the criminal-avatar is a Tony Montanya-like Italian American mafioso named Tommy Vercetti; *San Andreas* (2004) is set in a West Coast city, and the avatar-criminal is an African American gangsta named Carl Johnson, modeled on a character from the 1991 movie *Boyz N the Hood; GTA IV*'s *Liberty City* (2008) is a replica of New York, and

the digital lawbreaker is eastern European immigrant Niko Bellic—
a *Godfather* prototype. What the gamer does with the stolen automobile is a matter of choice, but violence and chaos seem unavoidable.
As in real life, the automobile is itself a weapon, a force for violence
and destruction. Digital cars, set aflame by assault rifle fire or Molotov
cocktail, explode with drivers still inside; pedestrians are run over—
some bounce off the car's grill, others fly over the hood. When a driver
hits a random motorcyclist, however, the resulting crash is particularly
catastrophic: the motorcyclist is sent flying long and high distances
before death à la cement trauma. It's also important to note that in
GTA the automobile can serve more banal and logistical purposes: it
can be used to go to a fast food joint, to have sex with a prostitute, or
to complete illegal errands. In *GTA* the automobile serves many purposes, but theft, violence, crime, and destruction are at the heart of the
game's digital automobility.

Stealing a car in *GTA*'s digital world is a discommodious combination of reality and fantasy. Car theft in *GTA* is undemanding and
nearly always without consequence. With a player's click of a console
button, the thief-avatar casually opens the door to an unmanned car
or tosses the driver out of an already occupied car and motors away.
Unencumbered by drivers, locks, the Club, alarms, OnStar, security
cameras, or any other theft-prevention system, automobile theft in
GTA is effortlessly accomplished. The thief's deed therefore becomes
an everyday activity.[19] "You will," as one reviewer counseled, "steal
thousands of cars in the course of the game, driving each until you have
destroyed it or until you see one you like better."[20]

The automobile is strangely disposable in this world, and the thief is
incorrigible. Even if the car thief were apprehended, he faces no court
system and no prison time: the criminal-avatar, whether arrested or
killed, regenerates in a designated place on the digital map. The digital
map, the more authentic component of *GTA*, is an immense and open-
ended arena built to mirror major American cities. The thief-avatar
navigates the freeways, manufacturing districts, slums, and urban
neighborhoods of a faux New York or Los Angeles in a range of digital
car makes and models that mimic the models on the streets. In *Vice
City Stories* you can steal the *Patriot* (Hummer), in *San Andreas* you

can steal the *Elegant* (BMW), and in *GTA IV* you can steal the *Infernus* (Lamborghini).[21] Also at the gamer's disposal are motorcycles, tractors, forklifts, trucks, buses, helicopters, and airplanes. The stolen cars and other vehicles perform, in some crucial respects, like cars on the street. The digital cars leave skid marks on the road after a sharp turn; they incur broken windows and lose fenders in accidents; and car radios play stations with commercials and popular music.[22] The incredible details of the game, coupled with the freedom made possible with the ease of car theft, make *GTA* a digital terrain of geographic reality and mayhem-based fantasy.

In the decade or so after the release of *GTA III* in 2002, the games have been a lightning rod for controversy, centering on real-world violence. In 2002, two teens and a man in his twenties from Grand Rapids, Michigan, spent a night drinking beer and running down digital pedestrians with stolen automobiles while playing *GTA III* and then went out on a real drive and ran down a thirty-eight-year-old man on a bicycle, stomped on him and punched him, and finally returned home to play the game.[23] The automobile, whether used as a weapon or as the innocent victim's conveyance, was the fulcrum of violence in real-world *GTA* incidents. *GTA* automobile theft entered reality when in 2003 Devin Moore, eighteen years old at the time and inspired by *Vice City,* killed three men in a police precinct and then, in classic *GTA* fashion, fled the scene in a stolen squad car.[24]

Some politicians, fearing the effects of *GTA* on children, reacted to the seemingly *GTA*-inspired murder sprees by calling for a new video-game rating system that would prevent adolescents from purchasing the game. The adolescent mind, reform legislators argued, was not able to separate reality from fantasy. In 2002 Joe Baca, a Democrat from southern California, introduced the Protect Children from Video Game Sex and Violence Act of 2002, asking legislators, "Do you really want your kids assuming the role of a mass murderer or a carjacker while you are away at work?"[25] A game of mayhem intended for adults' enjoyment, it seemed, often ended up in the hands of adolescents. *GTA,* they believed, threatened the mental health of American children. In a *Today Show* interview in 2004, famous activist-lawyer Jack Thompson called *GTA* a "murder and carjacking simulator."[26] Critics like Thompson also cite the sexualized aspect of *GTA:* the ability of the criminal-avatar, in a stolen car, to have sex with digital prostitutes. Critics were handed a smoking gun in 2005 when a secret sex scene, dubbed "Hot Coffee," was discovered in *GTA: San Andreas.* The code allowed the clothed CJ, after courtship, to have sex with a naked female-avatar. That year, New York senator Hillary Clinton launched a campaign on the national level to change *GTA*'s rating from M (Mature) to AO (Adults Only), with hopes that parents could more effectively protect their children.[27] Clinton, singling out *GTA* as the nation's most dangerous game, told the Kaiser Family Foundation that video games were a public health issue. "It is a little frustrating," she said, "when we have this data that demonstrates there is a clear public health connection between exposure to violence [in video games] and increased aggression that we have been as a society unable to come up with any adequate public response."[28] Despite the criticism from politicians and lawyers, *GTA* continued to sell hundreds of thousands of copies at fifty to sixty dollars a unit—a considerable percentage of them probably purchased by adolescents. Advocates for *GTA,* while admitting that these games were not intended for children, contended that the majority of gamers are adult men in their twenties and thirties, perfectly capable of separating fantasy and reality. They also pointed out that *GTA* was appealing because of actual game play and the expansive urban-action environment—not just violence. *GTA,* a game with automobile theft and automobile-inspired violence at its center, was defended as an adult stress-reliever.

Politicians and the game's apologists can't turn to academics, cultural critics, or technologists for straightforward answers because, unsurprisingly, they too disagree on the meaning of *GTA.* Journalist Steve Johnson and University of Wisconsin education theorist Paul Gee argue that games like *GTA* can be effective educational tools and also provide players with alternative social models. Johnson believes that gamers, motivated by rewards, learn how to perform complicated digital tasks—and therefore learn to decide, choose, and prioritize. "It's not what you're thinking about when you play the game," he writes. "It's the way you're thinking that matters."[29] Being a successful digital criminal is an effective learning exercise. Games like *GTA* can, for the better, challenge any singular definition of goodness. In a video game's world, Paul Gee writes, "what counts as being or doing good is determined by a character's own goals, purposes, or values, as these are shared with a particular social group to which he or she belongs."[30] The automobile thief and criminal in *GTA*, therefore, subscribes to the values of his community and acts on them. Performing the tasks necessary to win the game and learning the values of another community, Gee and Johnson believe, are effective pedagogy.

But writer Damon Brown, in his *Porn and Pong: How "Grand Theft Auto," "Tomb Raider," and Other Sexy Games Changed Our Culture* (2008), sees a darker side to the role *GTA* plays in American culture. The game, he argues, played *the* crucial role in desensitizing American media to digital porn and violence.[31] Brown believes that in the past twenty years, video games went from mirroring popular culture to setting important cultural trends; *GTA*—with the *Hot Coffee* incident and high-volume sales even in the face of severe criticism—was *the* hinge in this process of cultural transformation. Brown's argument is provocative, but he supports it with conveniently drawn conclusions. To Brown's credit, though, numerous scholars make the case that, despite Rockstar's design of this game as a critique of American violence and commercialism, *GTA* reinforces and promotes violence, racism, and sexism.[32] In a popular essay, engineer and ethicist Simon Penny argues that video games with gunplay, because they embody aspects of violent game play, teach players to blur reality and simulation. Therefore, games have the "potential to build behaviors that can exist without or separate from, and possibly contrary to, rational argument or ideology."[33] Penny suggests that games like *GTA* can train killers. The academic debate, like the political and moral one, is conflicted. But Rockstar continues to sell millions of units of *Grand Theft Auto*.

THE HACKER AS A CAR THIEF

Electrons are at the core of digital images and contemporary antitheft systems, and they are easily manipulated by those with special knowledge. At the highest intellectual and technical levels of auto thievery today, the criminals involved can be considered hackers. Since 2005, researchers at Johns Hopkins University, the University of California–San Diego, and the University of Washington have conducted laboratory experiments followed by practical demonstrations that show how easy it is to read Radio Frequency Identification (RFID) key codes and to take over an automobile's electronic control module without getting inside the car. Despite what both auto manufacturers and insurance companies don't want car owners to know, thieves do steal cars with RFID keys routinely, especially in Europe, where the technology has been in use for a longer time. The thieves first attach a microreader to a laptop computer. Then they capture radio signals, after positioning the equipment within a few feet of the target (a key if it is an active system, a car if a passive system). The microreader thus intercepts the transmissions sent out by an RFID key transponder, and the computer decrypts the code. Within twenty minutes a key can be made, and the thief is off! With RFID key kits now available because of consumer complaints concerning the cost and difficulty of obtaining replacement keys, the task can be even easier. Insurance companies still claim that the technology is uncrackable and that owners of stolen cars must be committing fraud, but that just places the onus on the consumer, an old strategy, as we have learned.[34]

In similar fashion, OnStar technology can also be easily defeated, either with a CD containing malware placed in the automobile's radio console, or outside the vehicle. Once done, the more than fifty computers in the car can be controlled, so that braking and acceleration systems can be taken over. And with the act accomplished, the software can be made to destroy itself, thus removing any evidence of tampering.

These far-reaching technologies have global implications in today's world of powerful underground drug and terrorist organizations. Their ability to defeat existing security measures requires innovative policing responses that transcend national borders. One indicator of this phenomenon is the rather bizarre instance of stolen American vehicles used as improvised explosive devices in Middle East conflicts.[35] With every economic shock, the market for stolen cars gains new currency.

CONCLUSION

Stealing the American Dream

In summary, with very few exceptions, American culture has characterized auto theft as a crime that is not terribly serious or important, unless violence accompanies the act or it enables its perpetrators to commit other crimes. Essentially, the car is disposable. If damaged or lost, the car is often easily replaced with another vehicle, perhaps one that's even better than the previous model. And, indeed, it is mass produced, made of uniform parts and in large numbers to effect economies of scale. But for many Americans, it is also an object of desire. And there lies the incongruity with parallel main currents in American life. For if Americans identify with their automobiles and have a love affair with a thing that evokes status and well-being, how can this object be seen in another light as something that can easily be replaced? If we are attached to this machine that has become a part of the family, how can we so easily say goodbye to it with no emotional remorse? Isn't the crime a personal violation? Some victims never feel that the recovered car is the same and can never feel comfortable in that vehicle afterward. Auto theft, then, can only be seen as a long-standing paradox in American life that is not easily resolved.

For generations, commentators, scholarly and otherwise, have waxed eloquent concerning the American love affair with the automobile, particularly as it applied to the mid-twentieth-century milieu, most specifically between the years 1955 and 1970. And some Americans did worship and love their cars in extreme ways, and still do to this day. But what do we mean by *some*? Many auto thieves connected sexuality and manhood to the automobile and were willing to take risks to drive one, even if it was not their own. As this study suggests, however, many more Americans had little affection for the car; it was merely the provider of basic transportation. Why else would so many car owners leave their keys in their cars, not caring whether they were stolen? Undoubtedly some owners were relieved to see the car gone. Or was the key left in the ignition a defiance of reality concerning how much trust was reasonable in America's urban neighborhoods? Why did so many cultural representations of auto theft portray the car as just another technological contrivance, to be taken and at times wrecked? To understand auto theft is to understand one glaring inconsistency in a postmodern world of things, symbols, and values. Perhaps we need to reexamine the authenticity of the American love affair with the automobile and work toward a more holistic view of the place of the automobile in American life.

As we completed our rather brief study, we discovered that what had started out as a topic on the margins of an examination of the automobile and American life has had far more to say about the center of that topic than we ever imagined. Just as the automobile has been a product of considerable creativity and technical ingenuity, auto thieves have exhibited those qualities too. The study has raised as many questions as it has answered, particularly in terms of institutional responses, cross-cultural relations, and human psychology and sociology. The problem of auto theft demanded a government response that overlapped jurisdictions and in doing so changed the nature of law enforcement practices and the limits of personal freedoms. The private industry with most at stake—the insurance industry—both lobbied for stricter law enforcement and pioneered scientific and technological measures to curb auto theft. Inventors working at both the center and the margins devised and constructed equipment that was ingenious and at the same time far from secure. Owners loved their cars but were often careless in protecting them. Architects designed structures to reduce crime, yet criminals largely ignored these concrete, stone, and wood fortifications. Crossing borders was, and remains, the car thief's best strategy. And for the most dedicated of these lawbreakers, taking a car was often just one of many illegal acts that in more recent times are intimately connected to gangs and drugs. As we moved forward in time, we examined a frayed society increasingly unraveling, tenuously held together by law enforcement and technology.

Ultimately, and no matter how far we may dig into the sources, we wonder whether something fundamental about us has changed with the coming of the automobile and the temptation to steal it. Greed, self centeredness, class and racial divisions, social perceptions, patterns of consumption, materialism, and the rise of a permissive society all figure in this mix. Have we become so addicted to speed and its various forms in our caffeinated society that personal mobility, status, and freedom outweigh sensibilities concerning our integrity and moral values?

Tables Summarizing Various U.S. Automobile Theft Crime Reports and Surveys, 1924–2010

TABLE 1.1

Vehicles stolen in Buffalo, New York, May 15–July 15, 1924

MAKE OF VEHICLE	NUMBER STOLEN	MAKE OF VEHICLE	NUMBER STOLEN
Auburn	2	Jewett	2
Buick	26	Jordan	5
Cadillac	5	Marmon	3
Chalmers	1	Maxwell	5
Chandler	1	Moon	1
Chevrolet	61	Nash	7
Cole	2	Oakland	3
Dodge	8	Oldsmobile	3
Dort	1	Overland	15
Durant	4	Packard	2
Elcar	1	Paige	1
Essex	1	Peerless	2
Ford	172	Star	1
Franklin	3	Stearns-Knight	2
Gardner	1	Studebaker	10
Haynes	4	Velie	2
Holmes	2	Wills St. Claire	5
Hudson	7	Willys Knight	5
Hupmobile	1		

Source: "Automobile Record Book for 1924," Buffalo, NY, in possession of author.

TABLE 1.2

Total auto thefts in towns over 25,000, known to police, 1930–1939

YEAR	AUTO THEFTS	YEAR	AUTO THEFTS	YEAR	AUTO THEFTS
1931	119,052	1934	88,420	1937	69,227
1932	100,604	1935	76,323	1938	58,490
1933	90,806	1936	66,973	1939	56,102

Source: FBI, Uniform Crime Reports, vol.16, no. 1 (Washington, DC: Government Printing Office, 1945), 12.

TABLE 2.1

Motor vehicle thefts, 1950–1980

YEAR	VEHICLES STOLEN	TOTAL CRIME
1950	82,866	1,790,030
1960	167,770	1,861,261
1970	921,400	5,568,200
1980	1,097,189	12,152,700

Sources: FBI, Uniform Crime Reports for the United States and Its Possessions (Washington, DC: Government Printing Office, 1950), 21:75, 105; Uniform Crime Reports for the United States (Washington, DC: Government Printing Office, 1960), 33, 88; Uniform Crime Reports for the United States (Washington, DC: Government Printing Office, 1970), 5, 28; Uniform Crime Reports for the United States (Washington, DC: Government Printing Office, 1980), 31, 37; Uniform Crime Reports for the United States (Washington, DC: Government Printing Office, 1990), 38, 50; Uniform Crime Reports for the United States (Washington, DC: Government Printing Office, 2000), 52; FBI, Crime in the United States 2009, www2.fbi.gov/ucr/cius2009/offenses/property_crime/motor_vehicle_theft.html (accessed May 24 2009).

TABLE 2.2
Makes of vehicles stolen

MAKE	NUMBER STOLEN	PERCENT OF TOTAL	MAKE	NUMBER STOLEN	PERCENT OF TOTAL
Buick	255	7.4	Imperial	8	0.2
Cadillac	96	2.5	Lincoln	23	0.6
Camaro	12	0.3	Mercury	60	1.6
Chevrolet	1,548	40.2	Mustang	100	2.6
Chrysler	40	1.0	Oldsmobile	210	5.4
Comet	18	0.5	Plymouth	124	3.2
Corvair	77	2.0	Pontiac	244	6.3
Corvette	76	3.0	Pontiac GTO	34	0.9
Dodge	84	2.2	Rambler	60	1.6
Falcon	28	0.7	Thunderbird	55	1.4
Ford	527	13.7	Volkswagen	57	1.5
Impala	88	2.3			

Source: House of Representatives, "US Department of Transportation Survey," in *Hearings before Subcommittee No. 5 of the Committee on the Judiciary, House of Representatives, H.R. 15215 and Related Bills,* March 6, 14, 1968, 90th Cong., 2nd sess. (Washington, DC: Government Printing Office, 1968), 35.

TABLE 2.3
Model years of automobiles stolen

MODEL YEAR	NUMBER STOLEN	PERCENT OF TOTAL
Before 1960	1,585	43.1
1960	247	6.4
1961	187	4.9
1962	302	7.8
1963	367	9.5
1964	425	11.0
1965	438	11.4
1966	325	8.4
1967	113	2.9

Source: "US Department of Transportation Survey," 35.

TABLE 2.4
Places of theft

PLACE OF THEFT	NUMBER STOLEN	PERCENT OF TOTAL
Public street—business	586	15.4
Public street—residential	1,212	29.5
Attended parking lot	144	3.8
Shopping center lot	161	4.2
Unattended parking lot	566	14.9
Public parking lot	163	4.3
Home garage or carport	177	4.7
Home driveway	332	8.7
New or used car lot	491	12.9
Car rental agency	62	1.6

Source: "US Department of Transportation Survey," 35.

TABLE 2.5
Purposes of theft

PURPOSE OF THEFT	NUMBER STOLEN	PERCENT OF TOTAL
Transportation	1,381	34.8
Joyriding	1,818	45.8
Sale	198	5
Sale of parts	125	3.1
In connection with crime	107	2.7
Escape	228	5.7

Source: "US Department of Transportation Survey," 35.

TABLE 2.6
Times of theft

TIME OF THEFT	NUMBER STOLEN	PERCENT OF TOTAL
Daylight	340	34.6
Dark	642	65.4

Source: "US Department of Transportation Survey," 35.

TABLE 2.7
Methods of entry

METHOD OF ENTRY	NUMBER STOLEN	PERCENT OF TOTAL
Car left unlocked	2,923	75.9
Forced	322	8.4
Key from another car	81	2.1
Key legally obtained	176	4.6
Key illegally obtained	351	9.1

Source: "US Department of Transportation Survey," 35.

TABLE 2.8
How a stolen car was started

METHOD OF STARTING	NUMBER STOLEN	PERCENT OF TOTAL
With key left in ignition	1,664	43.2
With key left in car but not in ignition	131	3.4
With Master, jiggler, or try-out key	182	4.7
Wires under hood jumped	152	3.9
Wires under dash crossed	245	6.4
Separate ignition system used	57	1.5
Ignition in "off" position	666	17.3
Tin foil used	68	1.8
Screwdriver or other tool used	93	2.4
Key legally obtained	207	5.4
Key illegally obtained	391	10.1

Source: "US Department of Transportation Survey," 35.

TABLE 2.9

Vehicle thefts and recoveries in the United States, 1956–1969

YEAR	NUMBER STOLEN	NUMBER RECOVERED	PERCENT RECOVERED
1956	263,700	246,050	93
1957	276,000	256,956	93
1958	282,800	260,176	92
1959	288,300	265,236	92
1960	321,400	295,688	92
1961	326,200	296,842	91
1962	356,100	320,490	90
1963	399,000	366,090	92
1964	463,000	412,070	89
1965	486,600	428,208	88
1966	557,000	459,360	82
1967	654,900	550,814	84
1968	777,800	668,908	86
1969	871,900	732,396	84

Source: Data compiled from *Uniform Crime Reports for the United States,* 1960, p. 88, Washington, DC; *Uniform Crime Reports for the United States,* 1970, p. 28, Washington, DC.

TABLE 2.10
Percentages of juveniles and adults arrested for motor-vehicle theft

YEAR	JUVENILES (UNDER 18)	ADULTS (18 AND OVER)	YEAR	JUVENILES (UNDER 18)	ADULTS (18 AND OVER)
1967	61.9	38.1	1973	56.4	43.6
1968	60.7	39.3	1974	55.1	45.0
1969	58.0	42.0	1975	54.6	45.4
1970	56.1	43.9	1976	52.6	47.4
1971	53.0	47.0	1977	53.0	47.0
1972	53.6	46.4	1978	50.6	49.4

Source: U.S. Senate, Subcommittee on Investigations of the Committee on Governmental Affairs, *Professional Motor Vehicle Theft and Chop Shops,* 96th Cong., 1st sess. (Washington, DC: Government Printing Office, 1980), 28.

TABLE 4.1

States with most motor vehicle thefts

STATE	1986	1996	2010	1996–2010 % CHANGE	1986–2010 % CHANGE
California	205,597	242,466	152,524	−37.1	−25.8
Texas	119,121	104,928	68,023	−35.2	−42.9
Florida	69,824	103,769	41,462	−60.0	−40.6
Georgia	26,264	46,215	30,305	−34.4	+15.4
Illinois	72,587	58,077	28,796	−50.4	−60.3
Michigan	72,024	62,930	26,875	−57.3	−62.7
Washington	14,037	28,893	25,729	−11.0	+83.3
Arizona	13,899	41,034	21,508	−47.6	+54.8
Ohio	40,396	45,528	21,508	−53.6	−47.7
New York	113,247	89,900	21,433	−77.3	−82.0
North Carolina	13,186	24,566	18,310	−25.5	+38.9
Maryland	24,334	36,083	18,051	−50.0	−25.8
Pennsylvania	42,130	49,690	16,669	−66.5	−60.4
Missouri	22,233	23,992	16,051	−33.1	−27.8
New Jersey	59,096	46,437	15,556	−66.5	−73.7
Tennessee	26,109	34,428	14,835	−56.9	−43.2
South Carolina	9,344	15,849	13,197	−16.7	+41.2
Indiana	18,027	24,817	13,118	−47.1	−27.2
National	1,224,127	1,395,192	737,142	−47.2	−39.8

Source: FBI, *Uniform Crime Reports,* 2011, Washington, DC.

TABLE 4.2

U.S. metropolitan statistical areas (MSA's) with the most motor-vehicle thefts, 2000

MSA	RANK	THEFTS	RATE PER 100,000
Chicago, IL	1	43,329	541.04
Detroit, MI	2	40,685	909.24
New York–Northern New Jersey–Long Island, NY-NJ-PA	3	39,970	458.76
Los Angeles–Long Beach, CA	4	39,531	423.70
Phoenix-Mesa, AZ	5	29,506	979.06
Houston, TX	6	25,416	633.66
Atlanta, GA	7	25,026	648.83
Washington, D.C.– MD-VA-WV	8	24,521	717.32
Philadelphia, PA-NJ	9	23,561	475.99
Miami, FL	10	20,812	956.59

Source: National Insurance Crime Bureau, "Standard Metropolitan Areas, Motor Vehicle Thefts," www.nicb.org/newsroom/nicb_campaigns/hot-spots/.

TABLE 4.3

U.S. MSA's with the most motor-vehicle thefts, 2010

MSA	RANK	THEFTS	RATE PER 100,000
Los Angeles–Long Beach–Santa Ana, CA	1	53,464	416.75
Chicago-Joliet-Naperville, IL-IN-WI	2	33,172	350.61
New York–Northern New Jersey–Long Island, NY-NJ-PA	3	29,189	154.46
Houston–Sugar Land–Baytown, TX	4	25,655	383.84
San Francisco–Oakland-Fremont, CA	5	22,617	521.68
Dallas–Fort Worth–Arlington, TX	6	21,964	344.68
Detroit-Warren-Livonia, MI	7	20,955	487.75
Atlanta–Sandy Springs–Marietta, GA	8	20,056	380.65
Miami–Fort Lauderdale–Pompano Beach, FL	9	19,572	351.72
Riverside–San Bernardino–Ontario, CA	10	18,269	432.42
Seattle-Tacoma-Bellevue, WA	11	16,192	470.72
Washington-Arlington-Alexandria, DC-VA-MD	12	15,559	278.73
San Diego–Carlsbad–San Marcos, CA	13	14,279	461.31
Phoenix-Mesa-Glendale, AZ	14	13,566	323.55
Philadelphia-Camden-Wilmington, PA-NJ-DE-MD	15	12,805	214.66

Source: National Insurance Crime Bureau, "Standard Metropolitan Areas, Motor Vehicle Thefts," www.nicb.org/newsroom/nicb_campaigns/hot-spots/.

TABLE 4.4

U.S. MSA's with the highest motor vehicle theft rates, 2000

MSA NAME	RANK	THEFTS	RATE PER 100,000
Phoenix-Mesa, AZ	1	29,506	812.40
Miami, FL	2	20,812	956.59
Detroit, MI	3	40,685	909.24
Jersey City, NJ	4	4,502	814.37
Tacoma, WA	5	5,565	807.92
Las Vegas, NV-AZ	6	10,971	794.37
Fresno, CA	7	6,897	783.90
Seattle-Bellevue-Everett, WA	8	18,242	781.26
Jackson, MS	9	3,191	737.55
Flint, MI	10	3,184	728.20

Source: National Insurance Crime Bureau, "Standard Metropolitan Areas, Motor Vehicle Thefts," www.nicb.org/newsroom/nicb_campaigns/hot-spots/.

TABLE 4.5

U.S. MSA's with the highest motor vehicle theft rates, 2010

MSA NAME	RANK	THEFTS	RATE PER 100,000
Fresno, CA	1	7,559	812.40
Modesto, CA	2	3,876	753.81
Bakersfield-Delano, CA	3	5,623	669.70
Spokane, WA	4	2,763	586.35
Vallejo-Fairfield, CA	5	2,392	578.69
Sacramento-Arden-Arcade-Roseville, CA	6	11,881	552.83
Stockton, CA	7	3,779	551.43
Visalia-Porterville, CA	8	2,409	544.80
San Francisco–Oakland-Fremont, CA	9	22,617	521.68
Yakima, WA	10	1,266	520.49

Source: National Insurance Crime Bureau, "Standard Metropolitan Areas, Motor Vehicle Thefts," www.nicb.org/newsroom/nicb_campaigns/hot-spots/.

TABLE 4.6

Motor-vehicle theft rates in neighborhoods

NEIGHBORHOOD	MOTOR-VEHICLE THEFTS PER 1,000 POPULATION	CHANCE OF HAVING YOUR VEHICLE STOLEN*
West Commerce Street, Dallas, TX	223.77	1 in 4
Lubertha Johnson Park, Las Vegas, NV	164.76	1 in 6
Peralta Villa, Oakland, CA	137.64	1 in 7
Dolittle Park, Las Vegas, NV	130.74	1 in 8
D Street, Las Vegas, NV	122.29	1 in 8
First Ward, Charlotte, NC	103.54	1 in 10
Astor Avenue, Commerce, CA	95.33	1 in 10
Cabrillo Road, San Jose, CA	95.05	1 in 11
Triangle, Milwaukee, WI	94.96	1 in 11
Jordan Downs, Los Angeles, CA	92.60	1 in 11

Source: "10 Worst Neighborhoods for Car Theft," February 14, 2011, www.dailyfinance.com /2011/02/14/10-worst-neighborhoods-for-car-theft/?a_dgi=aols (accessed July 9, 2012).

* If you live in this neighborhood for at least one year.

TABLE 4.7
Felony arrests of adults and juveniles for motor-vehicle theft, California, 1999–2010

	1999	2000	2001	2002	2003	2004
Total arrests	19,728	21,879	25,299	27,084	30,064	30,731
Adults						
Number	13,200	15,317	18,510	20,526	23,696	24,657
% of total	66.9	70.0	73.2	75.8	78.8	80.2
Juveniles						
Number	6,528	6,562	6,789	6,558	6,368	6,074
% of total	33.1	30.0	26.8	24.2	21.2	19.8

	2005	2006	2007	2008	2009	2010
Total arrests	30,717	27,927	22,582	17,010	14,245	13,091
Adults						
Number	24,818	22,503	18,069	13,596	11,297	10,804
% of total	80.8	80.6	80.0	79.9	79.3	82.5
Juveniles						
Number	5,899	5,424	4,513	3,414	2,948	2,287
% of total	19.2	19.4	20.0	20.1	20.7	7.5

Source: State of California, Department of Justice, "Crime in California," 2004 and 2010, Criminal Justice Statistics Center, Sacramento, CA, table 21.

Note: Motor vehicle includes autos, trucks, motorcycles, and other vehicles.

TABLE 4.8

Felony arrests for motor-vehicle theft by age group of arrestee, California, 2004 and 2010

TOTAL NUMBER	UNDER 18	18–19	20–29	30–39	40 AND OVER
2004 30,731	6,074	4,023	12,289	5,425	2,920
Percent of total	19.8	13.1	40.0	17.7	9.5
2010 13,091	2,287	1,562	4,767	2,623	1,852
Percent of total	17.5	11.9	36.4	20.0	14.1

Source: State of California, "Crime in California," table 32.

TABLE 4.9

Felony arrests for motor-vehicle theft by gender, California, 2004 and 2010

CATEGORY	2004	2010
Total number	30,731	13,091
Males	24,961	10,404
Percent	81.2	79.5
Females	5,770	2,687
Percent	18.8	20.5

Source: State of California, "Crime in California," table 31.

TABLE 5.1

NICB nontreaty process for recovery of stolen vehicles from abroad

1. Vehicle seized by foreign government.
2. Information relative to seizure received.
3. Vehicle identification number edited and corrected, as necessary.
4. Database inquiries made.
5. Response by NICB to foreign government inquiry (if any).
6. If no stolen record located, inquiry indexed for future reference.
7. If stolen with no member company interest, originating law enforcement and FBI notified.
8. If stolen with member company interest:
    a. Originating law enforcement agency and FBI notified.
    b. Certified copy of title obtained from member company.
    c. Certified copy of registration obtained from Department of Motor Vehicles.
    d. Certified copy of police report obtained from police department.
    e. Power of attorney prepared in English and language of seizing government.
    f. All documents translated into language of seizing government or, when required, translations secured in country.
    g. Apostille or certification of notaries secured from secretary of state's office.
    h. Documents submitted to seizing government's embassy or consulate within the United States for legalization.
    i. File forwarded to American embassy in seizing country.
    j. Foreign ministry of seizing government petitioned by American embassy for release.
    k. Signatures and legalization of foreign embassy/consular personnel in the United States certified by foreign ministry.
    l. Documents filed with seizing government or court.
    m. Vehicle released by foreign government.
    n. Possession taken and vehicle returned to United States.
    o. Member company notified upon return.
    p. Originating law enforcement agency notified to cancel theft record.

Sources: National Insurance Crime Bureau, *Upclose: International Operations,* September–October, 1999, 3; Annette Villarreal, supervisor, Texas Department of Public Safety, Border Auto Theft Information Center, personal communication, 2011.

Note: As outlined in the text, when the United States does not have a treaty on the recovery of stolen cars from a country, the process can be extremely difficult and tenuous, often resulting in nonrecovery.

Notes

INTRODUCTION. PARK AT YOUR OWN RISK

1. Jeffrey T. Schnapp, "Driven," *Qui Parle* 13 (Fall–Winter 2001): 138.

2. Sarah S. Lochlann Jain, "'Dangerous Instrumentality': The Bystander as a Subject in Automobility," *Cultural Anthropology* 19 (February 2004): 61.

3. Guillermo Giucci, *The Cultural Life of the Automobile: Roads to Modernity*, trans. Anne Mayagoitia and Debra Nagao (Austin: University of Texas Press, 2012), xi–xiv; John Urry, *Sociology beyond Societies: Mobilities for the Twenty-First Century* (London: Routledge, 2000).

4. James J. Flink and John B. Rae, leading automotive historians of the past generation, do not specifically mention auto theft. Professor Flink, however, did touch on broader themes cogent to the story. In his *America Adopts the Automobile, 1895–1910* (Cambridge, MA: MIT Press, 1970), chaps. 4 and 6, Flink addressed issues of government response and regulation that are directly related to important thematic currents in this study. As it turns out, patterns related to auto-theft measures established at the birth of the automobile generally stayed on course for more than a half century, with unsatisfactory results. Ashleigh Brilliant, *The Great Car Craze: How Southern California Collided with the Automobile in the 1920s* (Santa Barbara, CA: Woodbridge Press, 1989), 42–45, 119. Brilliant's work, largely overlooked for several decades before it was published in 1989, foreshadowed a transition that is currently taking place in the historiography of the automobile in America. Social and cultural history is increasingly supplementing the largely descriptive scholarship that is still favored by many auto history buffs.

5. Jurg Gerber and Martin Killas, "The Transnationalization of Historically Local Crime: Auto Theft in Western Europe and Russia Markets," *European Journal of Crime, Criminal Law, and Criminal Justice* 11 (2003): 215–26; Philip Gounev and Tihomir Bezlov, "From the Economy of Deficit to the Black Market: Car Theft and Trafficking in Bulgaria," *Trends in Organized Crime* 11 (2008): 410–29; Ragavan Chitra, "Why Auto Theft Is Going Global," *U.S. News & World Report* 126 (June 14, 1999): 16.

CHAPTER 1. "STOP, THIEF!"

1. *Horseless Age*, February 6, 1901, 37.

2. *Horseless Age*, May 7, 1903, 42.

3. "Locking Devices," *Horseless Age*, January 9, 1901, 19.

4. *Horseless Age*, February 6, 1901, 37.

5. Ibid., 19.

6. Ibid.

7. On the "chauffeur problem," see Kevin Borg, *Auto Mechanics* (Baltimore: Johns Hopkins University Press, 2008), 13–30. Edwin G. Klein's *The Stolen Automobile* (New York: Lenz & Reicker, 1919) features a thief dressed as a chauffeur in an unlikely plot that ends in the recovery of the car and a chance romance.

8. *Horseless Age* 12, no. 3 (1903): xviii.

9. U.S. Patent 928,824, issued July 20, 1909.

10. *National Automobile Theft Bureau: 75th Anniversary, 1912–1987* (n.p.: NATB, 1987), 6.

11. Fred J. Sauter, *The Origin of the National Automobile Theft Bureau* (n.p., 1949), 3. On the history of the NATB, see *Articles of Association of the Automobile Underwriters Detective Bureau* (n.p., n.d. [1917?]).

12. "Allege That Police Work with Automobile Thieves," *Horseless Age*, April 22, 1914, 619.

13. "Auto Insurance Growing in Favor," *New York Times*, April 24, 1910, XX5. On the topic of auto insurance, see Robert Riegel, "Automobile Insurance Rates," *Journal of Political Economy* 25 (June 1917): 561–79; H. P. Stellwagen, "Automobile Insurance," *Annals of the American Academy of Political and Social Science* 130 (March 1927): 154–62.

14. On the early history of the NATB, see *National Automobile Theft Bureau*.

15. Sauter, *National Automobile Theft Bureau*, 6. See *Constitution and Contract Membership of the National Automobile Theft Bureau* (n.p., 1928).

16. *National Automobile Theft Bureau*, 22.

17. "Two under Arrest Name Auto Thieves," *New York Times*, January 24, 1914, 2.

18. For an excellent discussion on the history of car keys, see Michael Lamm, "Are Car Keys Obsolete," *American Heritage Invention and Technology* 23 (Summer 2008): 7.

19. "Get $2,000 Payroll, Flee in Victim's Car," *New York Times*, September 27, 1925, 9.

20. Roy Lewis, "Watch Your Car," *Outing* 70 (May 1917): 170 (quote), 168; "The All-Conquering Auto Thief and a Proposed Quietus for Him," *Literary Digest* 64 (February 7, 1920): 111–15, with reference to Alexander C. Johnston's article in *Munsey's Magazine* (New York, 1920); "More than a Quarter of a Million Cars Stolen Each Year," *Travel*, October 1929, 46. See also William G. Shepard, "I Wonder Who's Driving Her Now?" *Colliers* 80 (July 23, 1927): 14. According to the *Oxford English Dictionary*, "joy ride" was first used in a 1909 article in the *New York Evening Post* reporting on a city ordinance that was passed to stop city officers from taking "joy rides."

21. For a comprehensive description of the auto-theft problem in the years immediately after World War I, see *Automobile Protective and Information Bureau, Annual Report* (n.p., 1921). Also see the pamphlet *A Growing Crime!* (n.p., n.d. [1923?]).

22. *Joyrider* as a term evoking irresponsibility and reckless disregard for others is briefly discussed in Peter D. Norton, "Street Rivals: Jaywalking and the Invention of the Motor Age Street," *Technology and Culture* 48 (2007): 342. On juvenile delinquency in the period, see Christopher Thale, "Cops and Kids: Policing Juvenile Delinquency in Urban America, 1890–1940," *Journal of Social History* 40 (Summer 2007): 1024–26; D. J. S. Morris, "American Juvenile Delinquency," *Journal of American Studies* 6 (December 1972): 337–40; Bill Bush, "The Rediscovery of Juvenile Delinquency," *Journal of the Gilded and Progressive Era* 5 (October 2006): 393–402.

23. Henry Barrett Chamberlain, "The Proposed Illinois Bureau of Criminal Records and Statistics," *Journal of the American Institute of Criminal Law and Criminology* 13 (February 1922): 522. Allegedly, the police moved one-third as fast as the criminals they chased.

24. Ellen C. Potter, "Spectacular Aspects of Crime in Relation to the Crime Wave," *Annals of the American Academy of Political and Social Science* 125 (May 1926): 12. Potter noted, "The automobile has added its spectacular element to causes for arrest in Philadelphia by approximately 10 percent. Assault and battery by the good, old-fashioned human fist lacks some of the elements which make the same offense by automobile a new story and more than 8,800 arrests were made in 1925 out of a total of 137,263."

25. George C. Henderson, *Keys to Crookdom* (New York: D. Appleton, 1924), 28–29.

26. W. S. Jennings, "Jeffersonville Reformatory, City of Renewed Hope, Where Indiana Keeps Some of Her Misfits," *Indiana Farmers' Guide* 33 (April 30, 1921): 4.

27. "Confessions of an Auto Thief," *Your Car,* June 1925, 34, 36.

28. William J. Davis, "Stolen Automobile Investigations," *Journal of Automobile Investigations* 28 (January–February 1938): 721.

29. Federal Bureau of Investigation, "Lawless Years: 1921–1933," www.fbi.gov /libref/historic/history/lawless.htm (accessed May 17, 2008).

30. Alexander Johnson, "Stop Thief!" *Country Life,* June 1919, 72.

31. Bennet Mead, "Police Statistics," *Annals of the American Academy of Political and Social Science* 146 (November 1929): 94; Arthur Evans Wood, "A Study of Arrests in Detroit, 1913 to 1919," *Journal of the American Institute of Criminal Law and Criminology* 21 (August 1930): 99.

32. Wood, "Arrests in Detroit," 94; Scott Bottles, *Los Angeles and the Automobile: The Making of a Modern City* (Berkeley: University of California Press, 1987), 92.

33. J. B. Thomas, *Conspicuous Depredation: Automobile Theft in Los Angeles, 1904–1987* (n.p.: Office of the Attorney General, California Department of Justice, Division of Law Enforcement, Criminal Identification and Information Branch, Bureau of Criminal Statistics and Special Services, March 1990).

34. *Theft of Automobiles,* Report 312, U.S. House of Representatives, 66th Cong., 1st sess. (Washington, DC: Government Printing Office, 1919), 1; Johnson, "Stop Thief!" 72.

35. "Car Insurance Is Withheld from Chicago Negroes: Appaling [*sic*] Auto Theft Rate Makes Negro Districts Bigger Risks than More Refined Districts," *Kansas City (KS) Plaindealer,* July 7, 1933, 1, 4.

36. Dana Gatlin, "In Case of a Thief," *Colliers* 56 (January 8, 1919): 94. See also Malcolm B. Johnson, "How to Foil the Auto Thief," *Youth's Companion* 103 (February 1929): 82–83.

37. "Two Kinds of Motor Thefts—Real and Imagined," *Literary Review* 74 (August 26, 1922): 50; "How Safe Is Your Automobile?" *Popular Mechanics Magazine* 42 (October 1924): 529.

38. "Two Kinds of Motor Thefts," 50.

39. John Brennan, "Automobile Thefts," *American City,* December 1917, 565–67.

40. Floyd Clymer, *Historical Scrapbook No. 4* (Los Angeles: Clymer, 1947), 162.

41. Brennan, "Automobile Thefts," 565.

42. "How Safe Is Your Automobile?" 532.

43. Simplex Corporation, Chicago, "Simplex Theftproof Auto Lock: A Lock for Every Car," Trade Catalog Collection, Benson Ford Research Center, Dearborn, MI.

44. Ad for Hershey Coincidental Locks, *Saturday Evening Post,* May 19, 1928, 50. See Orville S. Hershey, "Automobile Lock," U.S. Patent 1,685,128, issued September 25, 1928; Orville S. Hershey, "Automobile Lock," U.S. Patent 1,694,506, issued December 11, 1928; E. J. Van Sickel, "Automobile Lock," U.S. Patent 1,730,396, issued October 8, 1929.

45. Edward P. Francis and George De Angeles, *The Early Ford V8 as Henry Built It* (South Lyon, MI: Motor Cities, 1982), 53.

46. Stephen F. Briggs, "Lock," U.S. Patent 1,826,649, issued October 6, 1931; Robert F. Mangine, "Examination of Steering Columns and Ignition Locks," in *Forensic Investigation of Stolen-Recovered and Other Crime-Related Vehicles,* ed. Eric Stauffer and Monica S. Bonfanti (Amsterdam: Elsevier, 2006), 227–58.

47. Wallace C. D. Cochran, *Opportunity: A Pamphlet Addressed to Capital, and Pointing the Way to the Establishment of a Gigantic New Business Enterprise, viz:— Wholesale Automobile Theft Insurance, with the Risk Eliminated,* October 9, 1922, Vertical File "Theft Prevention," National Automotive History Collection, Detroit Public Library.

48. FEDCO Number Plate Corporation, *Foiling the Auto Thief: The FEDCO System of Automobile Theft Prevention and Detection,* December 1927, Vertical File "Theft Protection," National Automotive History Collection, Detroit Public Library.

49. The relationship between Chrysler and FEDCO apparently ended after the 1926 model year. See Thomas S. LaMarre, "From Model B to Big Three: Chrysler's Amazing Ascent," *Automobile Quarterly* 32, no. 4 (1996): 25.

50. Arch Mandel, "The Automobile and the Police," *Annals of the American Academy of Political and Social Science* 116 (November 1924): 193; Mead, "Police Statistics," 91–93; E. Austin Baughman, "Protective Measures for the Automobile and Its Owner," *Annals of the American Academy of Political and Social Science* 116 (November 1924): 197.

51. "Checking Automobile Thefts as Massachusetts Does It," *Literary Digest,* October 9, 1920, 84; James E. Bulger, "Automobile Thefts," *Journal of Criminal Justice and Criminology* 23 (January–February 1933): 808.

52. Mandel, "Automobile and the Police," 193.

53. From the *Congressional Record* 58 (1919): 5471.

54. *National Automobile Theft Bureau,* 24–25. William R. Outerbridge, Joseph D. Lohman, Albert Wahl, and Robert M. Carter, Supplemental Report, "The Dyer Act Violators: A Typology of Federal Car Thieves," typescript, School of Criminology, University of California, January 1967, ii.

55. Automobile Protective and Information Bureau, *Annual Report, 1920–1921* (n.p., 1921).

56. Ernest M. Smith, "Services of the American Automobile Association," *Annals of the American Academy of Political and Social Science* 116 (November 1924): 273.

57. "U.S. Agents Prepare to Hunt Auto Thieves," 1919, clipping, Vertical File "Theft Prevention," National Automotive History Collection, Detroit Public Library; "Federal Courts Divided over Meaning of the Dyer Act," *Columbia Law Review* 57 (January 1957): 130–32. See also *Congressional Record* 58 (1919): 5470–78.

58. Record Group 60, General Records of the Department of Justice, Formerly Classified Subject Correspondence, 1910–1945, Class 26, Dyer Act. For example, see Mrs. T. E. Young to Director, April 4, 1933, box 1870, file no. 26, folder 3; A. J. Roberts to Homer S. Cummings, October 8, 1934, file no. 26, folder 4; Morris Rockway to President of the United States, January 9, 1937, file no. 26, folder 5; Antoinette Aleanor Paveligo to President Roosevelt, July 18, 1940, box 1871, HM 1991.

59. Leonard D. Savitz, "Automobile Theft," *Journal of Criminal Law, Criminology, and Police Science* 50 (July–August 1959): 132–43, quote on 132. See also Report 61, Committee on the Judiciary, Subcommittee to Investigate Juvenile Delinquency, U.S. Senate, 84th Cong., 1st sess. (Washington, DC: Government Printing Office, 1955).

60. Edward Rubin, "A Statistical Study of Federal Criminal Prosecutions," *Law and Contemporary Problems,* vol. 1, *Extending Federal Powers over Crime* (October 1934): 501.

61. Davis, "Stolen Automobile Investigations," 731–32.

62. The tables are all gathered in the appendix.

63. Automobile Protective and Information Bureau, *Annual Report,* 19–21.

64. *Literary Digest* 66 (July 17, 1920): 77.

65. Lewis, "Watch Your Car," 169. W. S. Rowe, Cleveland Chief of Police, estimated that in 1916 a majority of autos were stolen by joyriders.

66. "All-Conquering Auto Thief," 112.

67. "Two Kinds of Motor Thefts," 50.

68. David Wolcott, "The Cop Will Get You: The Police and Discretionary Juvenile Justice, 1890–1940," *Journal of Social History* 35 (Winter 2001): 358.

69. The official distinction was drawn in *Impson v. State,* 47 AZ, 573, 1930. See also William R. Outerbridge et al., "The Dyer Act Violators: A Typology of Car Thieves," School of Criminology, University of California, January 1967, iv–v; Dyer quoted in Jerome Hall, "Federal Anti-Theft Legislation," *Law and Contemporary Problems* 1 (October 1934): 428. See also "Undoing Dyer," *Time Magazine,* March 24, 1930, 13.

70. "All-Conquering Auto Thief," 112.

71. John Landesco, "The Life History of a Member of the '42' Gang," *Journal of Criminal Law and Criminology* 23 (March–April 1933): 964–98.

72. Ibid., 980–84.

73. Edward C. Crossman, "How Your Automobile May Be Stolen," *Illustrated World,* March 1917, 34; "Tricks of the Auto Thief," *Popular Mechanics Magazine,* May 1929, 772–74; "Joe Newell Recovered 630 Stolen Automobiles Last Year," *American*

Magazine, February 1925, 67; Davis, "Stolen Automobile Investigations," 732; Edwin Teale, "Auto Stealing Racket Now $50,000,000-a-Year Racket," *Popular Science Monthly,* January 1933, 13.

74. Atcheson L. Hench, "From the Vocabulary of Automobile Thieves," *American Speech* 5 (February 1930): 236–37.

75. See Thomas J. Courtney, "Hot Shorts," *Saturday Evening Post,* November 30, 1935, 12. See also Shepard, "Who's Driving Her Now?" 14.

76. Teale, "Auto Stealing Racket," 96.

77. "Newell Recovered Stolen Automobiles," 67.

78. "Prisoner Says Ring Stole Cars to Order," *New York Times,* September 18, 1930, 1.

79. Ibid.

80. "Auto Gang Seized; Guaranteed Loot," *New York Times,* January 13, 1932, 31.

81. "Seized as Car Theft Ring," *New York Times,* October 12, 1935, 3.

82. "Held as Auto Theft Ring," *New York Times,* October 13, 1935, 25.

83. "6 Arrests Expose Big Car Theft Ring," *New York Times,* July 24, 1937, 3.

84. "Big Car-Theft Ring Bared by O'Dwyer," *New York Times,* January 4, 1942, 42.

85. Fred J. Cook, "Organized Crime: The Strange Reluctance," in *Investigating the FBI,* ed. Pat Watters and Stephen Gillers (Garden City, NY: Doubleday, 1973), 140.

86. Thomas A. Reppetto, *Bringing Down the Mob: The War against the American Mafia* (New York: Henry Holt, 2006), 8–9.

87. J. Edgar Hoover, "Bla-Bla Blackman," *American Magazine,* September 1936, 32.

88. Mexico—Providing for the Recovery and Return of Stolen or Embezzled Motor Vehicles, Trailers, Airplanes, or the Component Parts of Any of Them: Message from the President of the United States Transmitting a Convention between the United States of America and the United Mexican States for the Recovery and Return of Stolen or Embezzled Motor Vehicles, Trailers, Airplanes, or the Component Parts of Any of Them, Signed at Mexico City on October 6, 1936 (Washington, DC: Government Printing Office, 1937). The Bush administration made similar treaties with Panama (2000), Honduras (2001), and Guatemala (2002).

89. Convention with Mexico Providing for the Recovery and Return of Stolen or Embezzled Motor Vehicles, Trailers, Airplanes, or the Component Parts of any of Them: Report to Accompany Executive A, 75th Cong., 1st sess. (Washington, DC: Government Printing Office, 1937).

90. "Let the Auto Thief Beware," *Illustrated World,* August 1919, 858; J. Edgar Hoover, "The National Division of Identification and Information," *American Journal of Police Science* 2 (May–June 1931): 241–51.

91. Melvin Purvis, *American Agent* (Garden City, NY: Doubleday, Doran, 1936), 27–28.

92. On radio, including both police radio and public programming, see Kathleen Battles, *Calling All Cars: Radio Dragnets and the Technology of Policing* (Minneapolis: University of Minnesota Press, 2010), esp. 157–76.

93. Letter, off. Hartung to H. Jennings, n.d. [November 1927?]; James W. Higgins, chief of police, Buffalo, Chairman, "Auto Theft Committee," Int. Ass'n of Chiefs of

Police to Harry Jennings, November 4, 1927, both in University of Toledo, Records, 1899–1937, Toledo (Ohio). Police Dept., Stolen Auto Record, MSS 162, file, Correspondence to C. of P. Harry Jennings, 1924–27.

94. Horst Katterwe, "Restoration of Serial Numbers," in Stauffer and Bonfanti, *Forensic Investigation,* 177–94.

95. "Catching Auto Thieves," *American City,* October 1936, 15.

96. Sterling Gleason, "Auto-Stealing Racket Smashed by New Methods," *Popular Science Monthly,* August 1934, 12–13.

97. "Science Fights Crime with New Inventions," *Science News Letter,* March 16, 1935, 164.

98. Gleason, "Racket Smashed by New Methods," 13.

CHAPTER 2. JUVENILE DELINQUENTS, HARDENED CRIMINALS, AND SOME INEFFECTUAL TECHNOLOGICAL SOLUTIONS (1941–1980)

1. David Bogen, "Trends in Juvenile Delinquency," *Federal Probation* 9 (1945): 25–28.

2. Milwaukee Metropolitan Crime Prevention Commission, *Here, Kid, Take My Car* (n.p., 1946); James W. Horan, executive secretary, Milwaukee Metropolitan Crime Commission, *Keys to Another Kingdom* (n.p., 1947).

3. Leonard D. Savitz, "Automobile Theft," *Journal of Criminal Law, Criminology, and Police Science* 50 (July–August 1959): 132.

4. The auto-theft statistics used here are admittedly flawed and can provide only a qualitative guide to this activity. Several critics have pointed out the limitations associated with the FBI Uniform Crime Reports (UCR), first published in 1930. For example, see Sophia M. Robison, "A Critical View of the Uniform Crime Reports," *Michigan Law Review* 64 (April 1966): 1031–54. In addition to the flawed compilation system employed in the UCR's early history, their numbers are suspect from the perspective of their use as a measure of social control, as in cases involving juveniles and minorities. Local abnormalities, including underreporting, dictate a careful use of this material.

5. Michael G. Maxfield and Ronald V. Clarke, *Understanding and Preventing Car Theft* (Monsey, NY: Criminal Justice Press, 2004), 233.

6. U.S. Department of Commerce, *Statistical Abstract of the United States,* 1950, 71st ed. (Washington, DC: Government Printing Office, 1950), 137.

7. Maxfield and Clarke, *Understanding and Preventing Car Theft,* 237.

8. Beverly N. W. Lee and Giannina P. Rikoski, *Vehicle Theft Prevention Strategies* (Washington, DC: Government Printing Office, 1984), 3.

9. J. Edgar Hoover, "Auto Theft Is Big Business," *Motor Trend* 4 (December 1952): 17–18, 42.

10. On juvenile delinquents, see Jason L. Hostlutler, "Kids, Cops, and Beboppers," *Wisconsin Magazine of History* 93 (Autumn 2009): 14–27; Jason J. B. Bornosky, "The Violent Years: Responses to Juvenile Crime in the 1950s," *Polity* 38 (July 2006): 314–44; Sophia M. Robison, "Juvenile Delinquency," *Current History* 52 (May 1967): 341–48; William Graebner, "The Containment of Juvenile Delinquency: Social

Engineering and American Youth Culture in the Postwar Era," *American Studies* 27 (Spring 1986): 81–97.

11. Savitz, "Automobile Theft," 132.

12. Statistics taken from ibid., 133–34.

13. William W. Wattenberg and James Balistrieri, "Automobile Theft: A Favored-Group Delinquency," *American Journal of Sociology* 57 (May 1952): 575–79.

14. Max Gunther, "The Shocking Facts on Auto Theft," *Popular Science* 171 (December 1957): 110.

15. Logan A. Hidy, "A Study of Automobile Theft and the Juvenile Involved," master's thesis, Ohio University, 1951.

16. Ibid., 85–94.

17. Erwin Schepses, "The Young Car Thief," *Journal of Criminal Law, Criminology, and Police Science* 50 (March–April 1960): 569; "Girls Demands Cause Car Thefts?" *Los Angeles Tribune,* April 1, 1960, 9.

18. Erwin Schepses, "Boys Who Steal Cars," *Federal Probation* 25 (1961): 56–62.

19. See James A. Hamlett Jr., "Week-End Chats," *Kansas City (KS) Plaindealer,* April 18, 1958, 1; "Newton Still Tops the City Crimewise . . . ," *Los Angeles Tribune,* August 7, 1959, 3.

20. "Parker Checkmated by His Own Figures on 'Negro Crime,'" *Los Angeles Tribune,* August 7, 1959, 3.

21. *Rebel without a Cause,* directed by Nicholas Ray, with performers James Dean, Natalie Wood, Sal Mineo, and Corey Allen, Warner Bros. Pictures, 1955; videocassette (VHS), 1997; Lawrence Fascella and Al Weisel, *Live Fast, Die Young: The Wild Ride of Making "Rebel without a Cause"* (New York: Touchstone, 2005). Peter Biskind, "Rebel without a Cause: Nicholas Ray in the Fifties," *Film Quarterly* 28 (Autumn 1974): 32–38.

22. *Young and Wild,* directed by William Witney, Republic Pictures, 1958; Netflix (accessed December 1, 2011).

23. Theodore Weesner, *The Car Thief* (New York: Random House, 1972).

24. Ibid., 6.

25. *Boy in Court,* directed by David H. Lion, Willard Pictures, 1940; *Teenagers on Trial,* written and supervised by Frances Dinsmoor, RKO Pathe Pictures, 1955; *Car Theft,* Bray-Mar Productions, 1956; *Moment of Decision,* directed by Sid Davis, Sid Davis Productions, 1961.

26. *Joy Ride: An Auto Theft,* directed by William Crain, with performers Kari Markussen, Victoria Paige Meyerink, Tim Palladino, and Val Thorgusen, Barr Films, 1976, www.youtube.com/watch?v=8ILX7dzPYuQ (accessed May 30, 2011).

27. National Automobile Theft Bureau, "Police in Two Major Cities Declare an All-Out War on Auto Thieves," *Action Report* 1 (September–October 1964): 1, 4.

28. Ibid., 110–11. On how to use a "Slim Jim" to open a door, see the film at eHow, www.ehow.com/video 2331906 slim-jim-instructions.html.

29. Robert S. Chilimidos, *Auto Theft Investigation* (Los Angeles: Legal Book, 1971), 109–16.

30. Ibid., 243.

31. *National Automobile Theft Bureau: 75th Anniversary, 1912–1987* (n.p.: NATB, 1987), 56.

32. New York State Senate Committee on Transportation, *National Workshop on Auto Theft Prevention, Compendium of Proceedings* (Albany: New York State Senate, 1979), 44.

33. Chilimidos, *Auto Theft Investigation*, 244.

34. David Barry, H. Steinberg, W. Preysnar, E. Perchonok, and J. Collard, *Preliminary Study of the Effectiveness of Auto Anti-Theft Devices*, prepared for the National Institute of Law Enforcement and Criminal Justice, Law Enforcement Assistance Administration, U.S. Department of Justice, October 1975 (Washington, DC: Government Printing Office, 1975).

35. *Bonnie and Clyde*, directed by Arthur Penn, with performers Warren Beatty and Faye Dunaway, Warner Bros.–Seven Arts, 1967. On Bonnie and Clyde, see Robert B. Toplin, *History by Hollywood: The Use and Abuse of the American Past* (Urbana: University of Illinois Press, 1996); Jeff Guinn, *Go Down Together: The True, Untold Story of Bonnie and Clyde* (New York: Simon & Schuster, 2009); Mark Harris, *Pictures at a Revolution: Five Movies and the Birth of the New Hollywood* (New York: Penguin Press HC, 2008). For a close reading of *Bonnie and Clyde*'s significance in the development of the "road movie," see David Laderman, *Driving Visions: Exploring the Road Movie* (Austin: University of Texas Press, 2002), 51. See also Jack Sargeant and Stephanie Watson, "Looking for Maps: Notes on the Road Movie as Genre," in *Lost Highways: An Illustrated History of Road Movies*, ed. Jack Sargeant and Stephanie Watson (London: Creation Books, 2000), 6–20; Lawrence L. Murray, "Hollywood, Nihilism, and the Youth Culture of the Sixties: *Bonnie and Clyde* (1967)," in *American History / American Film*, ed. John E. O'Conner and Martin A. Jackson (New York: Ungar, 1988), 237–45. For the development of populist notions of the gangster, see David E. Ruth, *Inventing the Public Enemy: The Gangster in American Culture, 1918–1934* (Chicago: University of Chicago Press, 1996); John McCarty, *Bullets over Hollywood: The American Gangster Picture from the Silents to the "Sopranos"* (New York: Da Capo Press, 2004).

36. E. John Long, "My Car Was Stolen," *Collier's* 123 (March 19, 1949): 58.

37. "Six Are Arrested as Car Theft Gang," *New York Times*, August 13, 1947, 3; "3 Here Accused as Car Theft Ring," *New York Times*, November 24, 1947, 46; "Police Breaking Up Big Car Theft Ring," *New York Times*, January 18, 1948, 45; "300 Stolen Cars Sent out of U.S.; 6 Seized Here in $500,000 Racket," *New York Times*, December 4, 1949, 1; "10 Held in Car Theft Ring," *New York Times*, July 30, 1951, 11; "4 Arrested in Jersey in Auto Theft Ring," *New York Times*, December 9, 1952, 30; "20 Plead Innocent in Auto Theft Case," *New York Times*, April 18, 1953, 22.

38. Clyde Edgerton, *The Bible Salesman* (New York: Little, Brown, 2008).

39. *Hot Cars*, directed by Don McDougall, with performers John Bromfield and Joi Lansing, Bel-Air Productions, Sunrise Pictures, distributed by United Artists, 1956.

40. For example, see Myrl E. Alexander, acting parole executive, to Honorable Emanuel Celler, May 3, 1939; Gabriel Vigorito to U.S. Board of Parole, Judge L. W. Wilson, October 26, 1939; Myrl E. Alexander, acting parole executive, to Royal E.

Copeland, U.S. Senate, April 14, 1939, all in RG 60, Notorious Offenders File, "Gabriel Vigorito," National Archives, College Park, MD.

41. Stephen Fox, *Blood and Power: Organized Crime in Twentieth Century America* (New York: William Morrow, 1989), quote on 147–48.

42. "'No. 1' Car Theft Ring Broken; 4 Seized Here," *New York Times,* September 25, 1953, 28.

43. Will Oursler, "Hot Car King," *American Weekly,* January 9, 1955, 7ff.

44. "17 Admit Car Theft Ring," *New York Times,* April 19, 1956, 20.

45. "3 Face U.S. Charges on Car Thefts Here," *New York Times,* July 18, 1959, 7.

46. "8 Sentenced in Car-Theft Ring; Accountant Is Jailed as Leader," *New York Times,* February 10, 1962, 33.

47. "Martinis Case Bares Big-Car Theft Ring," *New York Times,* August 14, 1963, 66; "Martinis Inquiry Urged on State," *New York Times,* August 15, 1963, 31; "Accused Detective Suspended," *New York Times,* September 11, 1963, 34; "Inquiry on Increase in Auto Lot Thefts to Be Sought Here," *New York Times,* July 6, 1966, 69; "A Car Theft Ring That Adds Extras to Order Is Found," *New York Times,* July 21, 1966, 28; "Suspect Seized as Leader of Car-Theft Ring Here," *New York Times,* November 30, 1967, 40; "10 Are Indicted Here for Stealing Autos," *New York Times,* November 15, 1967, 36; "Clerk Is Seized as 'Key Figure' in a $10 Million Car-Theft Ring," *New York Times,* November 26, 1968, 22; "27 Are Indicted as Members of Ring Accused of Stealing 1,000 Cars a Year," *New York Times,* December 1, 1968, 33; "Alleged Head of Car Theft Ring Surrenders and Is Arraigned," *New York Times,* December 11, 1968, 18; "Volkswagen-Theft Ring Broken Up Here, F.B.I. Says," *New York Times,* May 17, 1969, 16; "Car-Theft Ring Adds a Twist: Guarantee to Re-steal Lemons," *New York Times,* June 4, 1969, 49; "21 Arrested in 'Steal to Order' Car-Theft Ring," *New York Times,* January 29, 1970, 28; "F.B.I. Arrests 16, Seeks Eight Others in Car Theft Ring," *New York Times,* June 21, 1970, 63; "Police Raid Car Wrecking Yard and Break Up Stolen Auto Ring," *New York Times,* August 19, 1970, 35; "Big Car Theft Ring Smashed in Bronx," *New York Times,* September 24, 1970, 1.

48. *Professional Motor Vehicle Theft and Chop Shops,* Subcommittee on Investigations of the Committee on Governmental Affairs, U.S. Senate, 96th Cong., 1st sess. (Washington, DC: Government Printing Office, 1980), 28.

49. The Right to a Quiet Society, www.quiet.org (accessed August 16, 2013).

50. Victor Helman, "Automatic Burglar Alarm for Automobiles," U.S. Patent 2,687,518, issued August 24, 1954; Joseph Gregory Yurtz, "Automobile Theft Alarm," U.S. Patent 2,885,504, issued May 5, 1959; Charles E. Davis, "Auto Alarm System," U.S. Patent 3,614,734, issued October 19, 1971.

51. R. L. Winklepleck, "Stamp Out Auto Theft," *Popular Electronics* 26 (March 1967): 59–61.

52. "Radatron Auto Sentry Burglar Alarm," *Radio-Electronics* 44 (March 1973): 26.

53. "Truth about Auto Burglar Alarms," *Mechanix Illustrated* 69 (April 1973): 106, 108–9.

54. John T. Frye, "Electronics and Car Thievery," *Popular Electronics* 2 (October 1972): 88, 90–91. For a do-it-yourself alarm system, see Frank J. Dielsi, "Vehicle

Alarm System: Keyless System Offers Three Operational Modes," *Popular Electronics* 1 (June 1972): 51–53.

55. S. M. Gallager, "World's Most Thiefproof Car Lock," *Popular Mechanics* 133 (April 1979): 38–39.

56. *Automobile Theft Prevention Act of 1979, S. 1214, Hearing Before the Subcommittee on Criminal Justice, Committee on the Judiciary, April 14, 1980,* U.S. Senate, 96th Cong., 2nd sess. (Washington, DC: Government Printing Office, 1980), 11.

57. Chapman Industries Corp., *Want a Car Free? Steal One! (Or How to Keep Yours Safe from Thieves),* 1987; "CHAPMAN-LOK and Total Protection Security Systems Operate from Inside the Car," 1979, Trade Catalog Collection, Benson Ford Research Collection, Dearborn, MI.

58. Aaron Friedman, Aaron Naparstek, and Mateo Taussig-Rubbo, *Alarmingly Useless: The Case for Banning Car Alarms in New York City* (New York: Transportation Alternatives, March 21, 2003).

59. U.S. Department of Transportation, National Highway Safety Administration, *Auto Theft and Recovery: Effects of the Anti Car Theft Act of 1992 and the Motor Vehicle Theft Law Enforcement Act of 1984,* report to the Congress (Washington, DC: Government Printing Office, 1998), 87.

60. For example, according to former Los Angeles insurance adjuster Don Queen, during the 1980s and 1990s the insurance industry had informal connections with Mexican police chiefs to return stolen cars for a "ransom." Don Queen, interview by John Heitmann and Rebecca Morales, February 16, 2011.

CHAPTER 3. FROM THE PERSONAL GARAGE TO THE SURVEILLANCE SOCIETY

1. The term "fortress America" was used by Edward J. Blakely and Mary Gail Snyder in their book *Fortress America: Gated Communities in the United States* (Washington, DC: Brookings Institution, 1997), but it has since been applied more broadly to refer to the national response to immigration, terrorism, and a host of other issues.

2. For background on the development of the garage, see Leslie G. Goat, "Housing the Horseless Carriage: America's Early Private Garages," *Perspectives in Vernacular Architecture* 3 (1989): 62–72; J. B. Jackson, "The Domestication of the Garage," *Landscape* 20 (Winter 1976): 10–19; and Folke T. Kihlstedt, "The Automobile and the Transformation of the American House, 1910–1935," *Michigan Quarterly Review* 19 (1980): 555–70.

3. Holly Wahlberg, "A House for the Automobile: The Changing Garage," *Old House Journal,* July–August 1998, www.oldhouseonline.com/a-house-for-the -automobile/.

4. Gerald Lynton Kaufman, "The Best Way to House Your Car: Build the Attached Garage for Utility, Beauty, and Economy," *American Home* 1 (January 1929): 302–3.

5. Henry Humphrey, "House Your Car," *American Home* 6 (April 1931): 26–30; "Architectural Portfolio: Attached Garages," *American Home* 12 (August 1934): 156–59.

6. Walter F. Wheeler, "Housing the Automobile: Some of the Requirements of the Up-to-Date Garage," *House Beautiful* 57 (February 1925): 139–42.

7. Greville Rickard, "Garages—Attached, Semi-attached, and Detached," *House & Garden* 67 (June 1935): 36–37, 75–76, 76.

8. "Widely Separated Inventors Invent Garage Door Openers by Radio Impulses," *Popular Science*, February 1931, 32.

9. "Aids to Modern Living—Garage Doors," *Popular Science*, January 1946, 137.

10. "Don't Use Horse-and-Buggy Thinking in Planning Your Garage," *House & Garden* 95 (January 1949): 82–83, 87.

11. "After the war THE GARAGE *MUST* BE RE-DESIGNED," *House Beautiful* 86 (June 1944): 70.

12. "Ups and Downs of the Garage Door," History Channel website, www.history .com/videos/ups-and-downs-of-the-garage-door (accessed June 8, 2011).

13. Setha M. Low, "A Nation of Gated Communities," in *The Insecure American: How We Got Here and What We Should Do about It*, ed. Hugh Gusterson and Catherine Besteman (Berkeley: University of California Press, 2009), 33.

14. Edward J. Drew and Jeffrey M. McGuigan, "Prevention of Crime: An Overview of Gated Communities and Neighborhood Watch," 2, www.ifpo.org /articlebank'gatedcommunity.html.

15. Haya El Nasser, "Gated Communities More Popular, and Not Just for the Rich," *USA Today*, www.usatoday.com/new/nation/2002-12-15-gated-usat_x.htm (accessed December 15, 2002).

16. Sven Bislev, "Privatization of Security as Governance Problem: Gated Communities in the San Diego Region," *Alternatives* 29, no. 5 (November–December 2004): 599–618. The author writes in note 4: "The definition is mine, adapted and developed from E. J. Blakely and M. G. Snyder, *Fortress America* (Washington, DC: Brookings, 1997) p. 2."

17. Bislev, "Privatization of Security," 601.

18. Low, "Nation of Gated Communities"; El Nasser, "Gated Communities More Popular."

19. Drew and McGuigan, "Prevention of Crime"; Bislev, "Privatization of Security," writes in his note 6: "For two reasons there are no robust statistics about the extent of gating. First, gating and use of other security measures is partly a matter of degree: for example, residences may have gates without guards or they may use guards and not have gates; they may have gates around part of common facilities or around all of them, and so on. See Blakely and Snyder, *Fortress America*, note 4. Second, in San Diego, at least, registration of gates that block entrance to residences is done by the fire department, but that is the only precise data, and it does not record the size of particular communities."

20. Edward J. Blakely and Mary Gail Snyder, "Separate Places: Crime and Security in Gated Communities," in Blakely and Snyder, *Fortress America*, 45.

21. Edward J. Blakely and Mary Gail Snyder, "Fortress Communities: The Walling and Gating of American Suburbs," *Land Lines* 7, no. 5 (September 1995): 1, www.lincolninst.edu/pubs/537_Fortress-Communities.

22. Setha M. Low, "The Edge and the Center: Gated Communities and the Discourse of Urban Fear," *American Anthropologist* 103, no. 1 (March 2001): 45–48, 53.

23. For example, author Heitmann's University of Dayton colleague Edward Garten had his Dodge Neon stolen while he was at a city restaurant. Although the car was soon recovered, its interior was littered with bottles and cigarette butts. Garten, feeling that his person had been violated, immediately traded the car for another model.

24. Barry Glassner, *The Culture of Fear* (New York, Basic Books, 1999).

25. Low, "Edge and the Center," 53.

26. Mary Newsom, "Gated Doesn't Equal 'Safer,'" in *Charlotte Observer's* blog *The Naked City,* posted December 8, 2009, writing about a public appearance by police chief Rodney Monroe, http://marynewsom.blogspot.com/2009/12/study -gated-doesnt-equal-safer.html.

27. Chris E. McGoey, "Gated Communities: Access Control Issues," www.crimedoctor.com/gated.htm (accessed May 9, 2011).

28. Barbara Ehrenreich, "What's So Great about Gated Communities?" 1, as reported in the Huffington Post, www.huffingtonpost.com, posted December 3, 2007.

29. Jane Jacobs, *The Death and Life of Great American Cities* (New York: Random House, 1961); C. Ray Jeffery, *Crime Prevention through Environmental Design* (Beverly Hills, CA: Sage, 1971); Oscar Newman, *Defensible Space: Crime Prevention through Urban Design* (New York: Macmillan, 1972); Oscar Newman, *Design Guidelines for Creating Defensible Space,* U.S. Department of Justice, National Institute of Law Enforcement and Criminal Justice (Washington, DC: Government Printing Office, 1976).

30. Herbert Gans, *People and Plans* (n.p.: Basic Books, 1968).

31. Marissa P. Levy and Christine Tartaro, "Auto Theft: A Site-Survey and Analysis of Environmental Crime Factors in Atlantic City, NJ," *Security Journal* 23, no. 2 (April 2010): 75–94; Marissa P. Levy and Christine Tartaro, "Repeat Victimization: A Study of Auto Theft in Atlantic City Using the WALLS Variables to Measure Environmental Indicators," *Criminal Justice Policy Review* 21, no. 3 (September 2010): 296–318.

32. Robert A. Gardner, "Crime Prevention through Environmental Design," 1995, www.crimewise.com/library/cpted.html (accessed May 10, 2011); this is a revised article by the same name as one originally published in the April 1981 issue of *Security Management Magazine.*

33. Ronald V. Clarke and Pat Mayhew, "Parking Patterns and Car Theft Risks: Policy-Relevant Findings from the British Crime Survey," in *Crime Prevention Studies,* ed. Ronald V. Clarke (Monsey, NY: Criminal Justice Press, 1994), 3:91–107; Todd Keister, "Thefts of and from Cars on Residential Streets and Driveways," *Problem-Oriented Guides for Police Problem-Specific Guides, Series No. 46,* February 2007.

34. John A. Jakle and Keith A. Sculle, *Lots of Parking: Land Use in a Car Culture* (Charlottesville: University of Virginia Press, 2004), 148.

35. Ibid., chap. 6, notes 49 and 50; here Jakle and Sculle supply references on lighting and safety.

36. James M. Tien, Vincent F. O'Donnell, Arnold Barnett, and Pitu B. Mirchandani, *Street Lighting Projects: National Evaluation Program: Phase 1 Report* (Washington, DC: U.S. Department of Justice, National Institute of Law Enforcement and Criminal Justice System, January 1979).

37. Stephen Atkins, Sohail Husain, and Angele Storey, "The Influence of Street Lighting on Crime and Fear of Crime," Crime Prevention Unit Paper No. 28, Home Office, London, UK, 1991, iii.

38. David P. Farrington and Brandon C. Welsh, "Effects of Improved Street Lighting on Crime: A Systematic Review," Home Office Research Study 251, Home Office Research, Development and Statistics Directorate, London, UK, August 2002, i.

39. Brandon C. Welsh and David P. Farrington, *Improved Street Lighting and Crime Prevention* (Stockholm: Swedish National Council for Crime Prevention, 2007), 11, as reported in Ronald V. Clarke, "Improving Street Lighting to Reduce Crime in Residential Areas," *Problem-Oriented Guides for Police Response Guides, Series No. 8,* December 2008, 11.

40. Jerry Ratcliffe, "Video Surveillance of Public Places," *Problem-Oriented Guides for Police Response Guides, Series No. 4*, U.S. Department of Justice, Office of Community Oriented Policing Services, Washington, DC, December 2006.

41. Marcus Nieto, "Public Video Surveillance: Is It an Effective Crime Prevention Tool?" California Research Bureau, June 1997, 13, www.library.ca.gov/crb/97/05/crb97-005.pdf (accessed July 27, 2010).

42. Ibid.

43. See also Barry Poyner, "Situational Prevention in Two Parking Lot Facilities," *Security Journal* 2 (1992): 96–101; and Barry Poyner and Barry Webb, *Successful Crime Prevention: Case Studies* (London: Tavistock Institute of Human Relations, 1987).

44. Kay Sunter, "Protecting Car Parks, Their Vehicles and Customers," *Facilities* 12, no. 9 (1994): 25–27.

45. Nieto, "Public Video Surveillance," 20.

46. Ibid., 23.

47. Stewart Ain, "Car Thefts Increase Sharply at Train Station," *New York Times,* June 20, 1993. This article, in addition to explaining how the Freeport parking lot made security cameras a central part of its security strategy, shows the difficulty of constructing a camera system. An effort in Nassau (a nearby municipality) to post signs for cameras and build a security camera system for a parking lot fell through because of the high cost and because of legal trouble that came with a false sense of protection, since the cameras had to be rotated between parking lots. For an example of a city responding to theft from parking lots with the construction of camera surveillance systems, see "BA to Point Cameras at Car Thieves," *Tulsa World,* July 6, 2008.

48. San Diego Police Department, "Vehicle Security," San Diego Police Department website, www.sandiego.gov/police/prevention/autotheft.shtml (accessed July 17, 2011). In addition to CCTV, authorities have used digital cameras that snap photographs of license plates and traffic cameras. See "Police Bust Band of Car

Thieves," *Manassas (VA) Journal Messenger,* June 23, 2009; and "The Policemen Might Look Away, but the Cameras' Eyes Don't Blink; HIGH-TECH: Sarasota's Plate-Recognition System Nets an Arrest in Car Theft," *Sarasota (FL) Herald Tribune,* December 23, 2008.

49. Nieto, "Public Video Surveillance," 22.

50. Ibid., 16.

51. Ronald V. Clarke and Patricia M. Harris, "Auto Theft and Its Prevention," *Crime and Justice* 16 (1992): 35.

52. Nieto, "Public Video Surveillance," 25.

53. Rana Simpson, "Theft of and from Autos in Parking Facilities in Chula Vista, California: A Final Report to the U.S. Department of Justice, Office of Community Oriented Policing Services of the Field Applications of the Problem-Oriented Guides for Police Project," Chula Vista, August 2004.

54. Ibid., 48. "The crime index, as presorted annually by police agencies to the FBI, consists of the following crimes, called part 1 crime: aggravated assault; auto theft; burglary; larceny; rape; robbery; and homicide. Theft from auto is contained in larceny Part 1 category" (48n1).

55. Simpson, "Theft of and from Autos," 14.

56. Ibid., 49–50. On checking for auto thefts at the border, the report adds these comments: "There is a license plate reader on the U.S. side of the border as you cross to Mexico. The National Review Subcommittee found that the plate reader is sometimes down—out of service for months at a time. The plate reader is also foiled if there is plastic (even clear plastic) covering the plate or if the vehicle passes too quickly past the reader into Mexico. There is also a license plate reader upon entry into California from Mexico. Essentially, the plate readers are not antitheft devices[;] they are a system to record, sometimes inaccurately, the number of times and at what time a vehicle enters or exits the border" (49–50n20). "Members of the San Diego Regional Auto Theft Task Force informed us that in years past, local police agencies teamed with federal border agencies to conduct random stops of cars about to enter Mexico. These were highly labor intensive, created traffic jams, and produced few arrests for auto theft over the years" (49–50n21).

57. Ronald V. Clarke, "Theft of and from Cars in Parking Facilities: Guide No. 10," Center for Problem-Oriented Policing, Office of Community Oriented Policing Services Office, U.S. Department of Justice, 2002, 10–15.

58. Nieto, "Public Video Surveillance," 6, quoting the U.S. Fifth Circuit Court.

59. Ibid., 9.

CHAPTER 4. CAR THEFT IN THE ELECTRONIC AND DIGITAL AGE (1970s–PRESENT)

1. "Joseph" (not his real name), a car and motorcycle thief, interview by John Heitmann and Rebecca Morales, Chula Vista, CA, police station, February 9, 2011.

2. *Gone in Sixty Seconds,* directed by H. B. Halicki, with performers H. B. Halicki, Marion Busia, and Jerry Daugirda, H. B. Halicki Mercantile Co., 1974. Halicki directed two other films in what became a trilogy: *The Junkman,* Bci/Eclipse, 1982, DVD, 2003; *Deadline Auto Theft,* Halicki Films, 1983, DVD, 2003.

3. *Grand Theft Auto,* directed by Ron Howard, with performers Ron Howard and Nancy Morgan, Buena Vista Home Entertainment, 1977, DVD. See also Hal Marcovitz, *Ron Howard* (Philadelphia: Chelsea House, 2002), 39–50; Roger Corman and Jim Jerome, *How I Made a Hundred Movies in Hollywood and Never Lost a Dime* (New York: Random House, 1990).

4. *Corvette Summer,* directed by Matthew Robbins, with performers Mark Hamill and Annie Potts, Turner Entertainment, 1978, DVD.

5. *Breathless,* directed by James McBride with performers Richard Gere and Valerie Kaprisky, Miko Productions, 1983, DVD.

6. See John Heitmann, *The Automobile and American Life* (Jefferson, NC: McFarland, 2009), 155–58.

7. Joe Bonamassa, "Tennessee Plates," written by John Hiatt and Michael Stuart Porter, www.songlyrics.com/joe-bonamassa/Tennessee-plates-lyrics/ (accessed December 8, 2011).

8. Sting, "Stolen Car," www.azlyrics.com/lyrics/sting/stolencar.html (accessed January 18, 2012).

9. Bruce Springsteen, "Stolen Car," written by Bruce Springsteen, Columbia Records, 1980, www.metrolyrics.com/stolen-car-lyrics-bruce-springsteen.html (accessed November 11, 2012).

10. Beastie Boys, "Car Thief," written by Beastie Boys and the Dust Brothers, Columbia Records, 1989, www.lyricstime.com/beastie-boys-car-thief-lyrics.html (accessed November 11, 2012).

11. *Professional Motor Vehicle Theft and Chop Shops,* Permanent Subcommittee on Investigations of the Senate Committee on Governmental Affairs, 96th Cong., 1st sess. (Washington, DC: Government Printing Office, 1980), 63. On the hearings, see "The Car-Theft Boom," *Newsweek,* December 10, 1979, 100–101.

12. *Professional Motor Vehicle Theft,* 70–71.

13. Ibid., 155.

14. Ibid., 184.

15. Michael deCourcy Hinds, "Bills Seek to Mark Car Parts as a Means to Reduce Thefts," *New York Times,* June 13, 1982, 63.

16. *No Man's Land,* directed by Peter Werner, with performers Charlie Sheen, D. B. Sweeney, and Lara Harris, Orion Pictures, 1987; MGM Entertainment, 2003, DVD.

17. Charles Klaveness, "At Last, Free of Cars and Insurers' Delays," *New York Times,* July 2, 1980, C1.

18. Donald T. DiFrancesco, "Juvenile Crime Must Be Handled with Maturity," *New York Times,* November 16, 1980, NJ 38.

19. Donald T. DiFrancesco, "A Plan to Reduce Automobile Thefts," *New York Times,* August 15, 1982, 26.

20. Michigan Automobile Theft Prevention Authority, "2011 Annual Report to the Governor and Legislature of the State of Michigan," 10, www.michigan.gov /documents/msp/2011/Annual_Report_Final_Rev_1_26_12_1135_AM_374794_7.pdf.

21. Debbie Yamamoto, California Department of Justice, to Rebecca Morales, February 17, 2011.

22. A cautionary note: the prominence of Los Angeles is to some degree an artifact of the changing and dynamic definition of metropolitan statistical areas from one period to another.

23. "Auto Theft Rates: U.S. Cities with the Highest and Lowest Auto Theft Rates," 1, Auto Insurance Tips, www.autoinsurancetips.com/auto-theft-rates-cities-highest-lowest-auto-theft-rates (accessed February 7, 2012).

24. Leroy C. Gould, "The Changing Structure of Property Crime in an Affluent Society," *Social Forces* 48, no. 1 (September 1969): 50–59; Thomas A. Reppetto, "Crime Prevention and the Displacement Phenomenon," *Crime and Delinquency,* April 1976, 166–77; Paul C. Higgins and Gary L. Albrecht, "Cars and Kids: A Self-Report Study of Juvenile Auto Theft and Traffic Violations," *Sociology and Social Research* 66, no. 1 (October 1981): 29–41.

25. *Menace II Society,* directed by Albert Hughes and Allen Hughes, with performers Tyrin Turner and Larenz Tate, New Line Productions, 1993, 2009, DVD; *New Jersey Drive,* directed by Nick Gomez, with performers Sharron Corley, Gabriel Casseus, and Saul Stein, Universal Studios, 1995, 2005, DVD.

26. *Gone in Sixty Seconds,* directed by Dominic Sena, with performers Nicholas Cage, Angelina Jolie, and Giovanni Ribisi, Bci/Eclipse Music, 2005, DVD. See also the novel based on the screenplay: M. C. Bolin, *Gone in 60 Seconds* (New York: Hyperion, 2000).

27. See *One Man's Blog,* http://onemansblog.com/2007/01/03/bmws-laser-cut-key-system-defeated-easily/ (accessed January 17, 2011), which describes how one person with a $1,200 tool can thwart a BMW laser key system in seconds.

28. *Chop Shop,* directed by Ramin Bahrani, with performers Alejandro Polanco and Isamar Gonzales, Koch Lorber Films, 2007, DVD. See *New York Times* film review, February 27, 2008, http://movies.nytimes.com/2008/02/27/movies,27chop.html (accessed May 4, 2011).

29. Michael G. Maxfield and Ronald V. Clarke, *Understanding and Preventing Car Theft* (Monsey, NY: Criminal Justice Press, 2004), 245.

30. Conversation between Sergeant Steve Witte, Chula Vista Police Department, San Diego County Regional Auto Theft Task Force, and John Heitmann and Rebecca Morales, February 9, 2011.

31. For example, see the brochure *Construct,* for a mechanical transmission lock manufactured by PAMACK, Heidelberg 26, 01796 Dohma, Germany.

32. Bruce Weber, "Jim Winner, Developer of the Club Antitheft Device, Dies at 81," *New York Times,* September 17, 2010, A18.

33. Nick Ravo, "The Club," *New York Times,* August 23, 1992, V10.

34. The Krook-Lok has evolved from its simple initial design, and a modified design can be purchased today. See Krooklok Vehicle Security Products, www.krooklok.co.uk (accessed February 18, 2011).

35. K. Zaidener, "Anti-Theft Device for Road Vehicles," U.S. Patent 3,245,239, issued April 12, 1966.

36. L. S. Baker and W. A. Tabor, "Automobile Lock," U.S. Patent 1,364,539, issued January 4, 1921; F. M. Furber, "Automobile Lock," U.S. Patent 1,448,658, issued March 13, 1923.

37. Marc W. Tobias assigned to Winner International Corporation, "Automotive Steering Wheel Anti-Theft Device," U.S. Patent 5,277,042, issued January 11, 1994.

38. On the history of the LoJack, see Funding Universe, www.fundinguniverse .com/company-histories/LoJack-Corporation-Company-History.html. See also Norm Alster, "A Car Thief's Nemesis," *Forbes,* May 11, 1992, 124; Tom Greenwood, "LoJack Helps Police Recover Stolen Vehicle," *Detroit News,* November 17, 2000; John R. White, "LoJack: That's a Technological Version of Kojak," *Boston Globe,* December 23, 1984, A57.

39. William R. Reagan, "Auto Theft Detection System," U.S. Patent 4,177,466, issued December 4, 1979.

40. Ian Ayres and Steven D. Levitt, "Measuring Positive Externalities from Unobservable Victim Precaution: An Empirical Analysis of LoJack," 3–4, Working Paper 5928, 1997, National Bureau of Economic Research, Cambridge, MA.

41. Marco Gonzalez-Navarro, "Deterrence and Displacement in Auto Theft," October 15, 2008, Princeton University Center for Economic Policy Studies, www.princeton.edu/ceps/workingpapers/177gonzalez-navarro.pdf (accessed May 30, 2012).

42. John Marcus, *Sourcebook of Electronic Circuits* (New York: McGraw-Hill, 1968); William M. Silfer Jr., "Method and Apparatus for Locating the Position of a Vehicle," U.S. Patent 3,357,020, issued December 5, 1967; Herbert Huebscher, "Position Monitoring System," U.S. Patent 3, 474,460, issued October 21, 1969.

43. Mario W. Cardullo and William L. Parks III, "Transponder Apparatus and System," U.S. Patent 3,713,148, issued January 23, 1973; Charles A. Walton, "Electronic Recognition and Identification System," U.S. Patent 3,816,708, issued June 11, 1974.

44. Erik Sandvik, "The Role of Technology in Reducing Auto Theft," unpublished paper [1996?], Florida Department of Law Enforcement, www.fdle.state.fl.us /Content/getdoc/d92354c9-5e20-4f0c-842d-dea82b9f1f66c/Sandvik.aspx, accessed May 30, 2012.

45. William D. Treharne, "Method and Apparatus for Enhanced Vehicle Protection," U.S. Patent 5,684,339, issued November 4, 1997. See also Ronald J. Robinson, "Automotive Device for Inhibiting Engine Ignition," U.S. Patent 4,645,939, February 24, 1987; Yohichi Iijima, "Antitheft Apparatus for Automotive Vehicle," U.S. Patent 5,519,376, issued May 21, 1996; Kanwaljit S. Khangura, William D. Treharne, and Ronald G. Moore, "Method and Apparatus for an Automotive Security System . . . ," U.S. Patent, 5,539,260, issued July 23, 1996.

CHAPTER 5. MEXICO, THE UNITED STATES, AND INTERNATIONAL AUTO THEFT

1. These dates draw on and expand upon those proposed by Carlos Antonio Flores Pérez in "Organized Crime and Official Corruption in Mexico," in *Police and Public Security in Mexico,* ed. Robert A. Donnelly and David A. Shirk (San Diego: University Readers, 2009), 93–124.

2. Lawrence D. Taylor, "The Wild Frontier Moves South: U.S. Entrepreneurs and the Growth of Tijuana's Vice Industry, 1908–1935," *Journal of San Diego History, San Diego Historical Society Quarterly* 48, no. 3 (Summer 2002): 1–12, www.sandiego history.org/journal/2002-3/frontier.htm.

3. Ibid., 5.

4. "Mexican Officials Pledge Cooperation in Recovering Stolen Cars across Border," *El Paso Herald,* December 18, 1920, edition 1, 3.

5. "Mexican Governor Pledges Aid against Auto Thieves," *Los Angeles Times,* October 4, 1925, H3, in ProQuest Historical Newspapers, Los Angeles Times (1881–1988), www.proquest.com/en-US/ (hereafter ProQuest).

6. "Lower California Co-operates against Smugglers of Stolen Autos: Mexico Gives Search Right," *Los Angeles Times,* Automobile section, November 14, 1926, G1, in ProQuest.

7. "Net Closes on Car Thieves: Western States Closely Linked in Work of Recovering Stolen Automobiles," *Los Angeles Times,* October 21, 1928, F1, in ProQuest. This article states, "Arrangements to prevent thieves from running stolen cars across the border into Mexico were also strengthened."

8. Rebecca Morales and Frank Bonilla, "Restructuring and the New Inequality," in *Latinos in a Changing U.S. Economy: Comparative Perspectives on Growing Inequality,* ed. Rebecca Morales and Frank Bonilla (Newbury Park, CA: Sage, 1993), 18–19.

9. Fred J. Sauter, *The Origin of the National Automobile Theft Bureau,* agency brochure (n.p.: February 10, 1949), 7.

10. "Convention for the Recovery and Return of Stolen or Embezzled Motor Vehicles, Trailers, Airplanes or Component Parts of Any of Them, signed at Mexico City, October 6, 1936," 50 Stat., pt. 2, p. 1333; *Diario Oficial* 103, no. 20 (July 23, 1937), sec. 2, pt. 1, 1333.

11. Alona E. Evans, "Treaty Enforcement and the Supreme Court of Mexico," *American Journal of Comparative Law* 5, no. 2 (Spring 1956): 267–70.

12. Dyer Act Class 26, Litigation Files: May 1, 1937, RG 60, National Archives, College Park, MD.

13. "Huge Dope Smuggling Ring Disclosed to Grand Jury: Girl Tells of Racket Where Cars Are Traded for Mexico Narcotics," *Los Angeles Times,* February 25, 1953, 2, in ProQuest.

14. Richard Valdemar, "Grand Theft Auto and Gangs," *Police,* March 25, 2010, www.policemag.com (accessed February 23, 2011).

15. Ibid.

16. For example, see H. F. Moorhouse, *Driving Ambitions: An Analysis of the American Hot Rod Enthusiasm* (Manchester: Manchester University Press, 1991); David N. Lucsko, *The Business of Speed: The Hot Rod Industry in America* (Baltimore: Johns Hopkins University Press, 2008).

17. C. Ronald Huff, *Comparing the Criminal Behavior of Youth Gangs and At-Risk Youths* (Washington, DC: National Institute of Justice Research in Brief, October 1998), 4.

18. Luis Astorga and David A. Shirk, "Drug Trafficking Organizations and Counter-Drug Strategies in the U.S.-Mexican Context," in *Shared Responsibility: U.S.-Mexico Policy Options for Confronting Organized Crime,* ed. Eric L. Olson, David A. Shirk, and Andrew Selee (Washington, DC: Woodrow Wilson International Center for Scholars; San Diego, CA: Trans-Border Institute, University of San Diego, 2010), 33.

19. "Plan to Cut Flow of Stolen Vehicles to Mexico Falters," *Los Angeles Times,* OC_A1, in ProQuest.

20. Tom Gorman and Robert Montemayor, "Car Theft 'Whitewash' on Mexican Side Feared: U.S. Probe of Massive Scheme Points to Involvement by High-Level Officials," *Los Angeles Times,* July 17, 1981, SD_A1, in ProQuest.

21. "8 Sentenced in International Theft Ring," *Los Angeles Times,* December 8, 1981, SD_A12; "Theft Ring Figure Gets 5-Year Term," *Los Angeles Times,* SD_A10, both in ProQuest.

22. Mark Forster, "Stolen Cars: Mexican Agents Rode Along—More Details Emerge in Court Papers on Massive Vehicle Ring," *Los Angeles Times,* SD_A1; "Suspect's CIA Ties Stymie $8-Million Car Theft Case," *Los Angeles Times,* A24, both in ProQuest.

23. Ronald J. Ostrow, "Resignation of U.S. Attorney Sought: Kennedy Chastised by Officials for Role in Exposing CIA Source," *Los Angeles Times,* OC_A14, in ProQuest.

24. Richard C. Paddock, "Ex-Mexico Official to Sue for Libel in Car-Theft Case," *Los Angeles Times,* April 21, 1982, SD_A1, in ProQuest, 1923–Current File.

25. Robert Montemayor, "Nazar Haro Testifies before Grand Jury: Ex-Director of Mexican Security Agency Denies Any Connection to Car Theft Ring," *Los Angeles Times,* April 23, 1982, SD_A1; Robert Montemayor, "Ex-Mexico Aide Seized in Car Theft Conspiracy," *Los Angeles Times,* April 24, 1982, B1; Robert Montemayor, "FBI Arrests Nazar Haro as Car-Thefts Conspirator: As He Ends S.D. Testimony," *Los Angeles Times,* April 24, 1982, SD1, all in ProQuest, 1923–Current File.

26. Robert Montemayor, "Ex-Mexico Official Named in Stolen Vehicles Case Fails to Show Up in Court," *Los Angeles Times,* May 4, 1982, C1; Robert Montemayor, "Nazar Haro Jumps Bail on U.S. Charge: Warrant Issued in Car-Thefts Case for Mexico's Former Chief of Security," *Los Angeles Times,* May 4, 1982, SD_A1; "Mexico May Never Return Accused Criminal to U.S.," *Los Angeles Times,* March 17, 1984, SD_A1, all in ProQuest, 1923–Current File.

27. Doreen Weisenhaus, "5 San Diegans Named Judges, Including Fired U.S. Attorney Kennedy," *Los Angeles Times,* July 19, 1983, SD_A1, in ProQuest, 1923–Current File.

28. *Hearing Before the Committee on Foreign Relations, On Treaty Document 97-18—Convention between the United States of America and United Mexican States for the Recovery and Return of Stolen . . .* , U.S. Senate, 97th Cong., 2nd sess. (Washington, DC: Government Printing Office, 1982), 19; Claude W. Cook, *The Automobile Theft Investigator: A Learning and Reference Text for the Automobile Theft Investigator, the Police Supervisor, and the Student* (Springfield, IL: C. C. Thomas, 1987), 36.

29. Juan M. Vasquez, "Graft Becomes a Public Issue for Mexicans," *Los Angeles Times,* June 21, 1982, B1, in ProQuest, 1923–Current File.

30. Rebecca Morales, *Flexible Production: Restructuring of the International Automobile Industry* (Cambridge, UK: Polity Press, 1994), 131.

31. Michael V. Miller, "Vehicle Theft along the Texas-Mexico Border," *Journal of Borderlands Studies* 2, no. 2 (1987): 18.

32. Ibid.

33. Professor David Shirk, director, Trans-Border Institute, University of San Diego, interview by Rebecca Morales, March 20, 2012.

34. Miller, "Vehicle Theft," 29.

35. Sergeant Richard Valdemar, interview by Rebecca Morales, March 9, 2011.

36. Miller, "Vehicle Theft," 18–19.

37. Ibid., 30n6.

38. Ibid., 19.

39. Richard Valdemar, "Blood Brothers: Roots of a Cartel War," *Police,* December 29, 2010, www.policemag.com. He goes on to state: "The 2000 movie 'Traffic' depicted part of the bloody conflict between the Arellano Felix brothers' Tijuana Cartel and the Carrillo Fuentes brothers' Juarez Cartel."

40. Valdemar, interview.

41. The *2005 National Gang Threat Assessment* (p.v.) as contained in *Criminal Investigation,* by Wayne W. Bennett and Kären M. Hess, 8th ed. (Belmont, CA: Thomson Higher Education, 2007), 555.

42. Valdemar, interview.

43. Ibid.

44. Miller, "Vehicle Theft," 30n9.

45. Darlanne H. Mulmat, Cynthia Rienick, Roni Melton, and Susan Pernell, "Targeting Auto Theft with a Regional Task Force and Mapping Technology," November 15, 2000, p. 31, document no. 185356, San Diego Association of Governments.

46. Miller, "Vehicle Theft," 12–32, 18–19.

47. Rosalva Resendiz, "Taking Risks within the Constraints of Gender: Mexican-American Women as Professional Auto Thieves," *Social Science Journal* 38 (2001): 475–81, 478–79. See also Rosalva Resendiz and David M. Neal, "International Auto Theft: The Illegal Export of American Vehicles to Mexico," in *International Criminal Justice: Issues in a Global Perspective,* ed. Delbert Rounds (Needham Heights, MA: Allyn and Bacon, 2000), chap. 1.

48. The article "US Targets 2 Sons of Sinaloa Cartel Leader," *San Diego Union Tribune,* May 8, 2012, speaks of women in cartels:

> On Monday, authorities in the northern border state of Nuevo Leon announced they had captured the female leader of a local cell of the violent Zetas drug cartel who is suspected of ordering or participating in at least 20 murders in or around the northern city of Monterrey. . . . State security spokesman Jorge Domene said suspect Maria Guadalupe Jimenez Lopez is nicknamed "La Tosca," or "the Rough One." Domene said that among the half-dozen suspected members of her gang arrested with her were three men who worked as hired killers and lookouts, earning between 4,000 and 10,000 pesos ($300 to $750) per month. . . . Three other members of the gang detained on May 1 are women: a 49 year-old mother and her

two daughters, aged 18 and 30, who allegedly worked as lookouts and sold drugs at a bar. . . . Domene said Jimenez Lopez received payments to oversee drug sales, auto theft, kidnappings and murders of rival gang members. . . . Women have been active in Mexico's drugs cartels, and a few have allegedly taken on high-level responsibility in the gangs. But women are usually employed by the cartels as drug couriers, lookouts or minor dealers; it is unusual to find them as leaders of local enforcement squads.

49. Alain G. Barbier, "International Collaboration Through Interpol," in *Forensic Investigation of Stolen-Recovered and Other Crime-Related Vehicles,* ed. Eric Stauffer and Monica S. Bonfanti (Burlington, MA: Academic Press, 2006), chap. 22, pp. 557–58.

50. Insurance Information Institute, *Auto Theft,* October 2010, 5, www.iii.org /media/hottopics/insurance (accessed February 23, 2011); Joe Hughes, "In Twist, Stolen Cars Return: Vehicles Get Phony IDs in Mexico for Sale in U.S.," *San Diego Union Tribune,* July 27, 2006.

51. David Agren, "Mexican Auto Dealers Decry Cross-Border Black Market," *USA Today,* September 24, 2012.

52. Murray Page, "The Saga of Mexico's Used Car Dilemma Caused by NAFTA," www.BanderasNews.com (accessed March 16, 2012).

53. Miguel Timo-Shenkov Ramirez, "Mexico Senate Nationalizes Stolen Cars," *Laredo Morning Times,* December 29, 2000, 1.

54. Agren, "Mexican Auto Dealers."

55. Ibid.

56. Insurance Information Institute, *Auto Theft,* 5.

57. Mulmat et al., "Targeting Auto Theft," 11.

58. Ibid., 12.

59. Ibid., 13.

60. Automobile Burglary and Theft Prevention Authority (ABTPA), "Auto Burglary and Theft Prevention Authority 20 Years: 1991–2011" (2011), 2, www.txdmv.gov.

61. Ibid., 6.

62. Ibid., 13.

63. Ibid., 10.

64. Annette Villarreal, supervisor, Texas Department of Public Safety, Border Auto Theft Information Center, personal communication, 2011.

65. ABTPA, "Auto Burglary and Theft Prevention Authority," 10. When asked what constituted "new techniques," the BATIC declined to elaborate in order to avoid compromising its efforts.

66. Miller, "Vehicle Theft," 27.

67. Ibid.

68. National Insurance Crime Bureau (NICB), *Upclose: International Operations,* September–October, 1999, 3. For a representative summary by the NICB of the nontreaty process, see table 5.1.

69. According to Miller, "Vehicle Theft," "Detectives with the state judicial police typically receive a salary of about $180 per month, and little in the way of guns and

other personal equipment. The state likewise generally fails to provide them with vehicles, although Chihuahua began furnishing official units to those in Juárez in mid-1987, in large part due to the negative publicity surrounding the use of stolen ones" (31n16).

70. Kristina Davis, "45 Arrested Locally in Nationwide Gang Crackdown," *San Diego Union Tribune,* March 1, 2011.

71. Kristina Davis, "Three El Cajon Men Charged in Iraqi Auto Theft," *San Diego Union Tribune,* August 31, 2011.

72. Greg Terp, "Examination of Vehicles Involved in Terrorism," in *Forensic Investigation of Stolen-Recovered and Other Crime-Related Vehicles,* ed. Eric Stauffer and Monica S. Bonfanti (Burlington, MA: Academic Press, 2006), chap. 17.

73. Dennis Frias, California operations manager, OCRA, and former FEAR agent, interview by Rebecca Morales, March 21, 2012.

74. Ibid.

75. Christopher T. McDonald, "The Changing Face of Vehicle Theft," *Police Chief* 78 (July 2011): 40–45, 44.

76. Ibid., 44.

77. Chris Woodyard, "Auto Exports from U.S. on Rise," *USA Today,* March 7, 2011.

78. Ken Ellingwood and Tracy Wilkinson, "Drug Cartels' New Weaponry Means War: Narcotics Traffickers Are Acquiring Firepower More Appropriate to an Army—Including Grenade Launchers and Antitank Rockets—and the Police Are Feeling Outgunned," March 15, 2009, *Los Angeles Times,* www.latimes.com. Ellingwood and Wilkinson cite McDonald, "The Changing Face," 44, as the source of their quotation.

79. Frias, interview.

80. Ibid.

81. See "El mojado acaudalado los tigres del norte," www.youtube.com/watch ?V=2G5X9Xc4T4Y (accessed August 23, 2013).

CHAPTER 6. THE RECENT PAST

1. *Gran Torino,* directed by Clint Eastwood, Warner Bros., 2008, DVD; Kenneth Turan, "Clint Eastwood, at 78, Shows He's Still a Formidable Action Figure," *Los Angeles Times,* December 12, 2008, www.latimes.com/entertainment/news/reviews /la-et-torino12-2008dec12,0,2314630-story (accessed December 6, 2011).

2. John R. Quain, "Better Antitheft Technology, but Smarter Car Thieves," *New York Times,* July 9, 2010. See also Heith Copes and Michael Cherbonneau, "The Key to Auto Theft: Emerging Methods of Auto Theft from the Offenders' Perspective," *British Journal of Criminology* 46 (2006): 917–34; Stephen Mason, "Vehicle Remote Keyless Entry Systems and Engine Immobilizers: Do Not Believe the Insurer That This Technology Is Perfect," *Computer Law and Security Review* 28 (April 2012): 195–200.

3. OnStar, www.onstar.com/web/portal/sva.2010 (accessed December 2, 2010).

4. Corvette Forum, http://forums.corvetteforum/com/c6-206-discussion /2292901-disabling-onstar.html (accessed January 19, 2011).

5. Asia-Jammer, www.asia-jammer.com/gps.htm (accessed February 20, 2011).

6. For example, in 2011 a theft ring in Montreal, Quebec, stole several Toyota Rav4, Highlander, Venza, and Lexus models using a microchip programmer, microchip keys, and a computer to penetrate the vehicle antitheft systems. They then shipped the cars to Africa and the Middle East. See Max Harold, "Vehicles' Anti-Theft System Disabled," www.montrealgazette.com/Three+arrested+global +theft+ring/4766206/sto. . . . (accessed November 14, 2011).

7. EC2/Global B2B Marketplace, www.ec21.com/product-details/Car-Remote -Control-Blocker-4346793.html (accessed February 20, 2011).

8. ChinaJammer, http://chinajammer.com/rm02-lojack-4g-xm-radio-jammer .html (accessed February 20, 2011).

9. *The Go-Getter,* directed by Martin Hynes, with performers Lou Taylor Pucci and Zooey Deschanel, Peace Arch Home Entertainment, 2008, DVD.

10. Janet Evanovich, *Motor Mouth* (New York: HarperCollins, 2006), 30–31.

11. Pete Hautman, *How to Steal a Car* (New York: Scholastic Press, 2009).

12. Ibid., 90.

13. Ibid., 117.

14. Ibid., 190.

15. Seth Schiesel, "Grand Theft Auto: The Story Continues, as Gritty as Ever," *New York Times,* February 18, 2009. *GTA III* was the best-selling video game of 2001, and *Vice City* was the best-selling game of 2002. When *Grand Theft Auto IV* was released in 2008, it took in more than $500 million in its first week. For details and arguments on *GTA*'s cultural impact, see Irene Chien, "Moving Violations," *Film Quarterly* 62 (2008): 80–81; Sorya Murray, "High Art/Low Life: The Art of Playing 'Grand Theft Auto,'" *PAJ* 27 (2005): 91–98; Kiri Miller, "Grove Street Grim: *Grand Theft Auto* and Digital Folklore," *Journal of American Folklore* 121 (2008): 255–85; Nate Garrelts, ed., *The Meaning and Culture of Grand Theft Auto: Critical Essays* (Jefferson City, NC: McFarland, 2005).

16. See John Leland, "Bigger, Bolder, Faster, Weirder," *New York Times,* October 27, 2002, H1, H12; Matt Richtel, "For Gamers Craving Won't Quit," *New York Times,* April 29, 2008; "Grand Theft Auto Sales Top $500 Million," *New York Times,* May 7, 2008. *GTA III* was published in 2002, *GTA: Vice City* was published in 2003, *GTA: Liberty City Stories* was published in 2005, *GTA IV* was published in 2008, and *The Ballad of Gay Tony* and *The Lost and the Damned* were published in 2009.

17. For a brief history and details of the game, see Bill Loguidice and Matt Barton, *Vintage Games: An Insider Look at the "Grand Theft Auto," "Super Mario," and the Most Influential Games of All Time* (Burlington, MA: Focal Press, 2009), 105–22.

18. Tim Bogenn and Rick Babra, *"Grand Theft Auto San Andreas": Official Strategy Guide, Signature Series* (Indianapolis, IN: Brady Games, 2003), 44–45.

19. See Timothy J. Welsh, "Everyday Play: Cruising for Leisure in San Andreas," in Garrelts, *Meaning and Culture of Grand Theft Auto,* 138. Welsh, with insight, writes, "Carjacking is so unremarkable that the extensive list of statistics, which includes everything from number of girls dated to legitimate kills, does not include a stat for number of cars stolen. All of the excitement, challenge and freedom is not in stealing cars, but in what one does with them afterwards. Stealing cars in GTA is as everyday in San Andreas as opening a car door is in the lived world" (138).

20. Charles Herold, "Game Theory: Stealing Cars and Telling Stories," *New York Times,* January 10, 2002, G6.

21. For the makes and models of *GTA* cars, see Tim Bogenn, *Grand Theft Auto: Liberty City Stories Official Strategy Guide* (Indianapolis, IN: Brady Games, 2005), 16–22; Tim Bogenn, *Grand Theft Auto: Vice City Stories Official Strategy Guide* (Indianapolis, IN: Brady Games, 2002), 15–25.

22. On the radio in *Grand Theft Auto,* see Kiri Miller, "Jacking the Dial: Radio, Race, and Place in 'Grand Theft Auto,'" *Ethnomusicology* 51 (2007): 402–38. Music is central to the digital car-theft experience. One of Miller's interviewees told her, "Country music is more appropriate for a stolen pickup truck, or hardcore rap for a low-rider. And some of the more relaxing stations are better suited for long drives in the country, while others may be better suited for, y'know, doing a drive-by."

23. For details on Grand Rapids, see David G. Myers, *Exploring Psychology* (New York: Worth, 2005), 565. See Shira Chess, "Playing the Bad Guy: Grand Theft Auto in the Panopticon," in *Digital Gameplay: Essays on the Nexus of Game and Gamer,* ed. Nate Garrelts (Jefferson, NC: McFarland, 2005): 80–90. The year after, two boys who claimed to be copying *GTA* shot at vehicles on a highway near their Newport, Tennessee, home and killed a forty-five-year-old driver. On Newport, see Maxine Frith, "'Grand Theft Auto' Makers Sued over Teenage Killing," *Independent* (UK), September 18, 2003.

24. Rebecca Leung, "Can a Video Game Lead to Murder?" *CBS News,* February 11, 2009, www.cbsnews.com/stories/2005/03/04/60minutes/main678261.shtml (accessed July 9, 2011). In 2004 a group of young criminals from Oakland, California, went on a string of carjackings, murders, and robberies. One of the arrested youth told the police, "We played the game by day and lived the game by night." Kim Worthy, "How Violent Video Games Can Cultivate Real Youth Violence: Prosecutor Says Grand Theft Auto–San Andreas and Other Games Have Dramatic Impact on Young Players," *Michigan Chronicle* 69 (September 28–October 4, 2005): A1.

25. Quoted in David Kushner, "The Road to Ruin: How Grand Theft Auto Hit the Skids," *Wired Magazine,* March 9, 2007, www.wired.com/gaming/gamingreviews /news/2007/03/FF_160_rockstar (accessed May 23, 2011).

26. Numerous individuals and organizations have criticized *Grand Theft Auto* as a threat to children. In 2004 California assemblyman Leland Yee, D–San Francisco, introduced a bill that restricted the sale of video games to minors. When the bill encountered resistance in the state legislature, Yee said: "Here we have children playing these violent video games for long periods of time—shooting, burning, maiming—all of these heinous acts. I thought it was a slam-dunk bill . . . but all of a sudden, people are hesitant, wondering what is wrong with the current system." See "Violent Video Games under Fire in Assembly, Bill Banning Minors from Buying M-Rated Volumes Has Its Foes," *Sacramento Chronicle,* April 5, 2004. In 2008 Mothers Against Drunk Driving asked the Entertainment Software Ratings Board to reclassify *GTA IV* as "Adults Only" because the game included a drunk driving sequence. See "MADD Attacks 'Grand Theft Auto IV,'" *Associated Press,* May 1, 2008. Michael Bloomberg said, "[*GTA IV*] doesn't exactly teach the kind of things that you'd want to teach your kids. . . . [It teaches] children to kill." See "Rants Begin

against Grand Theft Auto IV," *GamePro,* May 3, 2008. In 2005 Michigan governor Jennifer Granholm said, "We should all be disturbed by the availability of these games. . . . It wasn't just a problem in one store or one county, and it wasn't just a problem in large cities or rural communities. Children across the state have access to games that depict graphic violence and sexual exploitation." See Worthy, "How Violent Video Games," A1.

27. "Clinton Urges Inquiry into Hidden Sex in Grand Theft Auto Game," *New York Times,* July 14, 2005, B3.

28. "Clinton Seeks Uniform Ratings in Entertainment for Children," *New York Times,* March 10, 2005, B5.

29. Steve Johnson, *Everything Bad Is Good for You: How Today's Popular Culture Is Actually Making Us Smarter* (New York: Riverhead Books, 2005), 41.

30. Paul Gee, *What Video Games Have to Teach Us about Learning and Literacy* (New York: Palgrave Macmillan, 2003), 141–42. See also Lawrence Kutner and Cheryl M. Olson, *Grand Theft Childhood: The Surprising Truth about Violent Video Games* (New York: Simon and Schuster, 2008).

31. Damon Brown, *Porn and Pong: How "Grand Theft Auto," "Tomb Raider," and Other Sexy Games Changed Our Culture* (Port Townsend, WA: Feral House, 2008).

32. See, for example, David Leonard, "Virtual Gangstas Coming to a Suburban House Near You: Demonization, Commodification, and Policing Blackness," 49–69; Dennis Redmond, "Grand Theft Video: Running and Gunning for the U.S. Empire," 104–14; and Laurie N. Taylor, "From Stompin' Mushrooms to Bustin' Heads: Grand Theft Auto III as Paradigm Shift," 115–26, all in Garrelts, *Meaning and Culture of Grand Theft Auto.*

33. Simon Penny, "Representation, Enaction, and the Ethics of Simulation," in *First Person: New Media as Story, Performance, and Game* (Cambridge, MA: MIT Press, 2004), 81.

34. Brad Stone, "Pinch My Ride," www.wired.com/archive/14.08/carkey_pr .html (accessed June 20, 2012); Karl Koscher, Alexei Czeshis, Franziska Roesner, Shwetak Patel, Tadayoshi Kohno, Stephen Checkoway, Damon McCoy, Brian Kantor, Danny Anderson, Hovav Shacham, and Stefan Savage, "Experimental Security Analysis of a Modern Automobile," *2010 IEEE Symposium on Security and Privacy,* http://autosec.org; Stephen Checkoway; Damon McCoy, Brian Kantor, Danny Anderson, Hovav Shacham, Karl Koscher, Alexei Czeshis, Franziska Roesner, and Tadayoshi Kohno, "Comprehensive Analyses of Automotive Attack Surfaces," *IEEE Symposium, 2011;* R. Boyle, "Proof of Concept Carshark Software, Hacks Car Computers, Shutting Down Brakes, Engines, and More," *Popular Science,* May 14, 2010, www.popsci.com/cars/article/2010-05/researches-hack-car-computer-shutting -down-brakes-engine-and-more (accessed June 12, 2012).

35. Greg Terp, "Examination of Vehicles Involved in Terrorism," in *Forensic Investigation of Stolen-Recovered and Other Crime-Related Vehicles,* ed. Eric Stauffer and Monica S. Bonfanti (Amsterdam: Elsevier, 2006), 433–55.

Essay on Sources

The personal anxieties and legal uncertainties associated with the Dyer Act and automobile theft—the drama of it all—are best found in the hundreds of boxes of litigation case files (1919-60) that are class 26 of Record Group 60, Department of Justice, housed at the National Archives located in College Park, Maryland. These papers, which remain to be exhaustively mined, provide a rich personal account. The College Park National Archives collections also have an extensive file in Record Group 129, Notorious Offender Files, on the "King of Auto Thieves," Gabriel Vigorito. On the state and local level, the University of Toledo Archives, located in the Carlson Library, has the City of Toledo auto-theft records (1919-29), which are especially significant since that city's police chief played a leading role in organizing national efforts to deter car theft during the 1920s. Other primary source materials drawn on for this study include a vertical file, "Theft Protection," found at Detroit's National Automotive History Collection that contains pamphlets on locks, alarms, and other novel devices marketed primarily during the 1920s. Ephemeral materials on a wide variety of alarms developed and sold during the 1970s and 1980s are available at the Benson Ford Archives in Dearborn, Michigan. And an "Automobile Record Book for 1924," Buffalo, New York, in the possession of author Heitmann, provides one example of the scale of the problem during the 1920s.

One can find literally hundreds of government documents on the topic of auto theft, including many authored by individuals working for regional, state, and federal government agencies. Of these, the most important are the following U.S. House of Representative and Senate Reports: *Juvenile Delinquency: Part 18: Auto Theft and Juvenile Delinquency* [microform]*: Hearings before the United States Senate Committee on the Judiciary, Subcommittee to Investigate Juvenile Delinquency,* 90th Cong., 1st sess., January 17, 18, 27; April 19, 20, 1967 (Government Printing Office, 1967); *Auto Theft Prevention Act of 1968: Hearings before the United States House Committee on the Judiciary, Subcommittee No. 5 (Judiciary),* 90th Cong., 2nd sess., March 6, 14, 1968 (Government Printing Office, 1968); *Motor Vehicle Theft Prevention Act of 1980: Report Together with Additional and Dissenting Views, House of Representatives* (Government Printing Office, 1980); *Anti-Car Theft Act of 1992: Hearings before the Subcommittee on Crime and Criminal Justice of the Committee on the Judiciary, House of Representatives,* 102nd Cong., 1st and 2nd sess., on H.R. 4542 . . . , December 9, 1991, and March 31, 1992 (Government Printing Office, 1992). In terms of the relationship between the United States and Mexico concerning the issue of automobile theft, relevant government documents include these: Convention with Mexico Providing for the Recovery and Return of Stolen or Embezzled Motor Vehicles, Trailers, Airplanes, or the Component Parts of Any of Them: Report to Accompany Executive A, 75th Cong., 1st sess. (Government Printing Office; U.S. General Accounting Office, 1937); U.S. Customs Service, *Efforts to Curtail the Exportation of Stolen Vehicles: Report to the Chairman, Permanent Subcommittee on Investigations, Committee on Governmental Affairs, U.S. Senate / United States General Accounting Office* (General Accounting Office, 1999). One important government document from Mexico is "Convention for the Recovery and Return of Stolen or Embezzled Motor Vehicles, Trailers, Airplanes, or Component Parts of Any of Them, signed at Mexico City, October 6, 1936," 50 Stat., pt. 2, *Diario Oficial* 20 (July 23, 1937): 1333.

Several government-sponsored studies also proved particularly useful in our work. Caroline Harlow Wolf's *Motor Vehicle Theft* (U.S. Dept. of Justice, 1988) is a valuable primer on auto theft as it occurred during the 1970s and 1980s. So is Beverly N. W. Lee and Giannina P. Rikoski's *Vehicle Theft Prevention Strategies* (U.S. Dept. of Justice, 1984).

These government sources and many more similar ones can be accessed as electronic documents from library and other websites. The availability of searchable databases has revolutionized the practice of historical research during the past decade. Useful databases include Google Scholar (especially valuable for accessing complete patent records); America: History & Life and History of Science, Technology & Medicine (for secondary sources); JSTOR (for scholarly and legal sources); African American Newspapers: New York Times Historical; and Readers' Guide Retrospective. The last three Internet sources in particular pointed us to innumerable contemporary references in the periodical literature on juvenile delinquents, professional auto thieves, the built environment, and a wide variety of technologies and strategies to deter auto theft.

The novel can easily become a dangerous source for a historian to draw from. Yet to ignore these sources would result in a cultural void. Often used to explore the latent meanings of auto theft and the many motives of those involved, fiction can be a treacherous bedfellow if not used with caution. With some trepidation, however, one can consult novels to texture these events in ways that government documents and the news media do not. Furthermore, and on a macroscopic level, during the past decade the explosion of popular fiction that centers on auto theft reflects Americans' fascination as well as uneasiness concerning the topic. With those caveats, perhaps the best single novel on car theft reflective of these tensions is Theodore Weesner's

The Car Thief (Random House, 1972). Recent fiction of note includes M. C. Bolin's *Gone in 60 Seconds* (Hyperion, 2000), Janet Evanovich's *Motor Mouth* (HarperCollins, 2006), Timothy Watt's *Grand Theft* (Putnam, 2003), Clyde Edgerton's *The Bible Salesman* (Little, Brown, 2008), and Pete Hautman's *How to Steal a Car* (Scholastic Press, 2009).

Using film poses several issues similar to those that pertain to novels, yet quite different methodological challenges. Among the films drawn on for this study are the following: *Bonnie and Clyde* (1967), *Corvette Summer* (1978), *Chop Shop* (2007), *Gone in Sixty Seconds* (two versions, 1974 and 2000), *Grand Theft Auto* (1977), *Gran Torino* (2008), *Hot Cars* (1956), *New Jersey Drive* (1995), *No Man's Land* (1987), *Rain Man* (1988), and *Teenagers on Trial* (1955). Monographs useful in interpreting these films include Joyce Appleby's *Liberalism and Republicanism in the Historical Imagination* (Harvard, 1992), Deborah Clarke's *Driving Women: Fiction and Automobile Culture in Twentieth-Century American Fiction* (Johns Hopkins, 2007), James Gilbert's *A Cycle of Outrage: American Reactions to the Juvenile Delinquent in the 1950s* (Oxford, 1986), Stephen Greenblatt's *Cultural Mobility: A Manifesto* (Cambridge, 2009), Grace Elizabeth Hale's *A Nation of Outsiders: How the White Middle-Class Fell in Love with Rebellion in Post-War America* (Oxford, 2011), Michael Kummel's *Manhood in America: A Cultural History* (Oxford, 2012), Paula J. Massood's *Black City Cinema: African American Urban Experiences in Film* (Temple, 2003), and Jeremy Packer's *Mobility without Mayhem: Safety, Cars and Citizenship* (Duke, 2008).

Works of academic scholarship directly linked to automobile history and characterized by the broadest level of analysis include, in addition to John Heitmann's *The Automobile and American Life* (McFarland, 2009), recent studies in the direction of social and cultural themes: David Gartman's *Auto Opium* (Routledge, 1995), Kevin Borg's *Auto Mechanics* (Johns Hopkins, 2008), Cotton Seiler's *Republic of Drivers* (Chicago, 2009), David Blanke's *Hell on Wheels* (Kansas, 2008), David N. Lucsko's *The Business of Speed* (Johns Hopkins, 2008), Peter Norton's *Fighting Traffic* (MIT, 2008), Kathleen Franz's *Tinkering: Consumers Reinvent the Early Automobile* (Pennsylvania, 2005), and Joseph Corn's *User Unfriendly: Consumer Struggles with Personal Technologies, from Clocks and Sewing Machines to Cars and Computers* (Johns Hopkins, 2011). What all of these scholars have in common is their use of the automobile as a handle for delving into the topics of class, power, and the confounding complexities of technology at the human interface. On another level, these studies demonstrate just how central the automobile was to the twentieth-century American experience.

On the history of crime in America, we found only a few direct references to auto theft. David Wolcott's article on juvenile delinquency in the *Journal of Social History* has only two paragraphs on the subject. There are occasional but significant references in Claire Bond Potter's *War on Crime: Bandits, G-Men, and the Politics of Mass Culture* (Rutgers, 1998). Among writers who focused on auto theft along the U.S.-Mexican border was Michael V. Miller, who wrote the essay "Vehicle Theft along the Texas-Mexico Border," published in the *Journal of Borderland Studies* in 1987, and Rosalva Resendiz, who coauthored the chapter "International Auto Theft: The Illegal Export of American Vehicles to Mexico" in Delbert Rounds's *International Criminal Justice: Issues in a Global Perspective* (Allyn and Bacon, 2000) and wrote the pathbreaking work "Taking Risks within the Constraints of Gender: Mexican-American Women as Professional Auto Thieves," which appeared in the *Social Science Journal* in 2001.

Potter's study is an examination of how, during the interwar years, J. Edgar Hoover's FBI employed scientific investigative methods, systematic evidence collection, and analysis to forge a federal police state that was connected to emerging capitalist institutions and associated social relations. She argues that during the New Deal 1930s, the state was under considerable pressure from "auto bandits," the likes of John Dillinger and Clyde Barrow, who stole cars before committing more serious crimes and then got away on America's newly constructed system of improved highways. These bandits were quite different from their urban ethnic gangster counterparts, since they were more often found in the vast stretches of Missouri, Oklahoma, and Kansas, crossing state lines and confounding local authorities. To check this sort of crime, Hoover systematically connected fingerprints, modus operandi, and physical descriptions to hunt down the public enemies, thus to restore order and confidence in government. Although Potter has bigger fish to fry than the common auto thief, she clearly recognizes the importance of the crime institutionally and socially during a discrete time in history. For Potter, auto theft served the FBI as an issue around which it could rebuild its reputation beginning in the mid-1920s. Furthermore, drawing on cross-disciplinary insights, Potter saw similarities between auto thieves and other perpetrators of property crimes, on one hand, and Eric Hobsbawm's precapitalist bandits as articulated in *Social Bandits and Primitive Rebels* (Free Press, 1960), on the other hand.

And while criminologists' studies on auto theft abounded in the period after 1970, the leading social science scholar in the field, Ronald V. Clarke, lamented as late as 1992 that he could not unearth a solitary academic book published in the English-speaking world during the previous two decades on the subject. However, in that seminal essay, Clarke and coauthor Patricia M.

Harris raised important questions that continue to perplex law enforcement authorities and insurance investigators. Ongoing issues include the thorny matter of comparisons between rural and urban auto-theft rates, interurban variations, contrasts in various property crime trends, variations in targeted brands, and national disparities.

While Clarke's work examining the contemporary scene is invaluable, Andrew A. Karmen's sociological study has also proved to be relevant to our scholarship. Writing in the wake of Ralph Nader's *Unsafe at Any Speed* (Grossman, 1965), Karmen argued that, as in the case of automobile safety, the industry deflected auto-theft responsibility away from manufacturers to car owners and drivers. In essence, drivers were damned and the car was spared. He further asked whether the automobile industry was guilty of deliberately manufacturing cars that were easily stolen because of design limitations in the vehicles' locks, windows, and electrical systems. Nader had accused industry in general of negligence and hubris; in this context, were automobile companies socially responsible? Karmen supported his claims of industry neglect by examining statements from representatives of the Automobile Manufacturers Association, the Chrysler Corporation, the Ford Motor Company, the General Motors Corporation, and the National Automobile Dealers Association. Consequently, Karmen leaned heavily on a 1967 statement by General Motors President James Roche, when he testified in a Senate hearing, that owner carefulness was far more significant a deterrent than the installation of safety devices. For Karmen, the onus was not on the owners, as Roche maintained, but on the Big Three for their lack of interest in installing even the simplest of antitheft devices on their vehicles.

Karmen's now thirty-year-old study directs a significant accusation at American car manufacturers. Did manufacturers, until recently, turn a blind eye to the problem of automobile theft? Was it in their best interest to manufacture cars that were easily broken into, then started with little difficulty and driven off—since the victim often would purchase a new vehicle even if the stolen car was later recovered? It was hard to continue a love affair with something that had been violated, and according to Karmen, the auto industry knew that. Yet Karmen based his accusations on a very limited number of documents, and as our story has unfolded, with considerable texture added, his compelling claim comes into serious question. However, his assertions cannot be totally refuted; their defense or rejection awaits further forays into the history of automobile theft. As long as General Motors documents are largely inaccessible, Ford Motor Company archives contain little after 1956, and a large portion of Chrysler's materials remain in an uncertain state, historians must play the cards dealt them.

Index